ADVANCE PRAISE FOR *MID-COURSE CORRECTION* HAS COME FROM BUSINESS, EDUCATION, AND COMMUNITY LEADERS, AS WELL AS FROM ACTIVISTS, AUTHORS AND ENVIRONMENTAL ORGANIZATIONS:

There's a lot of talk about the importance of leadership in persuading the business community to operate more sustainably. But there is very little real leadership around. Ray Anderson's story provides a rare and inspiring exception.

Jonathon Porritt
Forum for the Future, Great Britain

[Ray Anderson's] direct, practical, no nonsense vision for corporate sustainability within the environment is emblematic of a man determined to embrace responsibility for being part of the solution rather than part of the problem... American business needs more men like him, and so does the planet.

Bernadette Cozart, Founder
The Greening of Harlem Coalition

In an industrial context often dominated by the status quo and limited or vague in its intentions, [Ray Anderson's] heroic activities show his commitment to a legacy of abundance and hope for generations to come. He is redesigning commerce from the inside out, ... and all our children will benefit.

William McDonough, Dean School of Architecture, University of Virginia
Author of *The Hannover Principles*

This is indeed an epiphany of great insight and urgency. Ray Anderson has caught a vision that is transforming his company and can help save us humans from ecological suicide. ...I rejoice in his innovative leadership and hope that the corporate world will follow his example and act responsibly in defense of planet Earth.

Bishop L. Bevel Jones (retired)
United Methodist Church, Bishop in Residence, Emory University

If you long for a finer world—an equitable, compassionate and sustainable habitat for all that live—read this stirring book. It will fire your hope that God's gift of an imperilled planet may survive the human assault on its riches and enlist you in the work of leaving a legacy for the yet unborn.

Bennett J. Sims, President, The Institute for Servant Leadership
Episcopal Bishop Emeritus of Atlanta

Ray Anderson has written a book which is a delightful combination of head and heart. Not only does he tell his fellow business leaders how to increase profits while improving the global ecosystem. He also tells us how good it feels to do it.

Stephan Schmidheiny, Founder
World Business Council for Sustainable Development

Many talk about the need for environmental stewardship. Personal example is the greatest teacher, and Mr. Anderson's book is the first full account to date of the CEO as environmental steward.

William Futrell, President
Environmental Law Institute

The sustainability challenge is truly daunting. I very much admire Interface's vision and Ray Anderson's willingness to accept this challenge. Through the use of Ray's road map... the path to sustainability is clearer.

General Dennis J. Reimer

This inspiring account of [Ray Anderson's] efforts to make Interface environmentally sustainable proves that business can provide jobs, meet clients' needs, and make a profit while leaving the world a BETTER planet for the generations to come.

Michael Baldwin, Founder
The Marion Foundation

This is three stories in one: personal, corporate, and spiritual. Taken together they spell out the remarkable journey of a highly competitive businessman who... is making his billion-dollar corporation the model of an earth-friendly, people-respecting, profit-enhancing business. [He] demonstrates how corporations can and must build a sustainable future. Bravo!

Ed Skloot, Executive Director
Surdna Foundation

This book is not about the environment; it is about life, society, us. Ray's energy and commitment to a sustainable enterprise are contagious and have influenced my personal interest in sustainable technology. I highly recommend this book to all... who want to influence 'the next industrial revolution.'

Dr. Jean-Lou Chameau, Dean, College of Engineering
Georgia Institute of Technology

Ray Anderson's vision is far-reaching and bold. He is, for sure, a business-man with guts. Ray has declared that Interface would be the first name in [industrial] ecology, worldwide, through substance, not words. To me, that's leadership.

J. Harold Chandler, Chairman, President & CEO
Provident Companies, Inc.

Ray's personal and professional odyssey is an excellent example of how one person can make a difference. The fact that he has been able to achieve all that he has while making Interface a successful company is a clear indicator that economics and environment are mutually compatible.

John H. Adams, Executive Director
Natural Resources Defense Council

This book should be required reading for all who genuinely care about the kind of earth that we want to leave for future generations.

Howard H. Callaway, Chairman
Callaway Gardens

[Mid-Course Correction] is an impressive document that makes the point that business leaders must be at the front of the sustainability parade if the necessary goals are to be achieved.

Samuel C. Johnson, Chairman
S. C. Johnson & Son

This book is a powerful story about a man who finally merged his vocation and 'advocation' and wants to save the world.

Michael A. Volkema, President & CEO
Herman Miller, Inc.

Everyone who leads a company or wants to and everyone who cares about the relationship between business and the environment will benefit from reading this book. Anderson writes from the head and the heart. His book is extraordinary; it is practical and wise and delightful to read.

Robert H. Dunn, President & CEO
Business for Social Responsibility

Ray Anderson's vision...provides a road map to a sustainable future. The challenges and opportunities it poses for Interface and all corporations are both substantial and surmountable.

Fred Krupp, Executive Director
Environmental Defense Fund

...What we need today, more than anything else, is intelligent, passionate and proactive leaders in business, who are bold enough to start demonstrating that sustainability in business pays. Ray Anderson is one of those. In his eyes, sustainable development in business is not only necessary, it is attractive.

Dr. Karl-Henrik Robèrt, Chairman & Founder
The Natural Step

A must-read for any executive with the courage to step forward and make a difference for the survival of the planet and the generations who will succeed us on it.

Wayne Clough, President
Georgia Institute of Technology

Ray Anderson's 'green' message is distinctive. ... [He's] demonstrated not only sustainability, but also profitable sustainability—what he calls 'doing well by doing good.'

James S. Balloun, Chairman, President & CEO
National Service Industries

In Mid-Course Correction, *[Ray Anderson] describes how one person and one company can change an industry, even industrialism, towards sensible hands-on sustainable practices. Here, will and gumption combine with the extraordinary genius of a true leader to change the world for the better.*

Paul Hawken
Author of *The Ecology of Commerce*

In a compelling and convincing way, this captain of industry shares his conversion experience from plunderer of to caretaker and advocate for the environment. Ray Anderson invites—even implores us—to make the mid-course correction necessary to save Spaceship Earth.

F. Stuart Gulley, President
LaGrange College

Ray Anderson is a man with a vision, and the will and power to make that vision happen. His story, told simply and with passion, is an important one for all of us to reflect upon.

Stephen Viederman, President
Jessie Smith Noyes Foundation

Anderson here presents a thrilling account of his own eco-odyssey, a journey of great significance and interest at both personal and corporate levels.

Peter H. Raven, Director
Missouri Botanical Garden

Mid-Course Correction *is an exciting and readable story... told with a compelling passion. It is a must-read for any environmentalist, but even more so for any hardnosed, pragmatic, skeptical business person who cares about the bottom line.*

Elliott H. Levitas, Former U.S. Representative
Attorney, Kilpatrick Stockton

This is a story of hope, vision and courage. Read it, and think of the world our children might enjoy if other business leaders seize the future as Ray Anderson has. Read it, and envision a world in balance... Read it, and regain hope that such a world is possible. Read it for your own sake, and for your children's.

Jonathan Lash
CoChair, President's Council on Sustainable Development
President, World Resources Institute

This most important and exciting business book of the 1990s distills the wisdom of a master of practical profitability and moral enterprise.

Amory B. Lovins, Vice President, CFO & Director of Research
Rocky Mountain Institute

While most businesses race blindly down the road of growth and profit, a few think differently. They see their purpose as contributing to a healthier society, and are confident that if they do that well, they will make plenty of money. Read this book to learn how one company is doing it.

Peter M. Senge, Ph.D., The Society for Organizational Learning
Senior Lecturer, Massachusetts Institute of Technology

Ray Anderson describes the economic sense that can be made of environmental concern, but on a much higher level, he makes us realize that the standard view of running a business just for the approval of Wall Street is dead wrong. We all have a stake in Spaceship Earth. Are we going to be leaders?

Thomas R. Oliver, Chairman & CEO
Bass Hotels & Resorts

MID-COURSE CORRECTION

The Peregrinzilla Press

ATLANTA

TOWARD A SUSTAINABLE
ENTERPRISE:
THE INTERFACE MODEL

Mid-Course CORRECTION

Ray C. Anderson

Published by The Peregrinzilla Press, Atlanta, Georgia

Printed in the United States of America

ISBN 0-9645953-5-4

What use is a house if you
haven't got a tolerable planet to put it on?

HENRY DAVID THOREAU

Our vision is of a life-sustaining Earth.

PRESIDENT'S COUNCIL ON SUSTAINABLE DEVELOPMENT

SUSTAINABLE AMERICA, 1996

ACKNOWLEDGEMENTS

I begin by acknowledging my debt to the past, personified by my parents. I think they would have been pleased by reading this and would write off their portion of that debt.

This book, drawn from thoughts developed in the process of preparing and giving some 200 speeches, would not exist without the infinite patience of the people who have transcribed my inadequate literary attempts through uncounted iterations. Janet Amundsen, my administrative assistant, slugged through the drafting and redrafting of all my earliest speeches upon which the bulk of the book is based; Kaye Gordy filled in during Janet's leave of absence; and, most recently, Cindy Stringer labored through each "fine tuning" redraft. My sincerest thanks go to each of them.

I thank Jim Hartzfeld and Jennifer DuBose for their assistance in working out the schematics used in Chapter Five and for compiling the list of PLETSUS ideas in the Appendix.

I acknowledge my wife, Pat, for her loving endurance of my many nights away from home as I have crisscrossed North America and two oceans to deliver the speeches that have consumed my life for the last three years.

I acknowledge the special contribution of Paul Hawken, Melissa Gildersleeve, Joyce LaValle, John Picard, and Gordon Whitener, which is revealed near the end of Chapter Seven. (It reads best if you wait until then.)

I also acknowledge the contributions of two new friends who have come into my life in connection with my membership on the President's Council on Sustainable Development: Jonathan Lash and Timothy Wirth. Tim's comments after reading an early draft were very encouraging, especially in light of his position at the time as Undersecretary of State for Global Affairs and, currently, as President of Ted Turner's United Nations Foundation. Jonathan's encouragement, also after reading an early draft, to write a "bigger book," influenced me to expand the part about myself that also deals with the founding of my company, Interface, Inc. Jonathan, I am afraid it is still a little book, but thank you for pushing me.

I acknowledge with special thanks the suggestions offered by Mike Bertolucci, Bill Browning, Paul Hawken, Amory Lovins, Fred Krupp, and Peter Raven that helped me tighten up the factual accuracy of some of the technical discussions and the help early on from Phyllis Mueller in editing preliminary drafts and, at the end, for cleaning up after me.

In many ways most important of all, I acknowledge the manifold ways that the people of Interface, led by a group called the Peregrinzillas (they know who they are), are actually doing the things to change our company—the changes that give the words of this book whatever credibility they possess.

I conclude by acknowledging my debt to the future, personified by my grandchildren, and their children, and theirs. I hope that their world will be a bit better than it otherwise would have been.

Ray C. Anderson
July 1998
Atlanta, Georgia

This book is not terribly original. Nor is it written with any attempt at scientific rigor. I make no effort to prove anything. It is largely a synthesis of the works of others whose insights and convictions, not to mention their inspired thinking and writings, have shaped my new-found attitude toward Earth, its fragile ecology, and my haunting role in its devastation. It is also a statement of what I have come to believe, but my education goes on. Among those who have inspired me directly are John Picard, Paul Hawken, Daniel Quinn, Bill McDonough, Bernadette Cozart, David Brower, Karl-Henrik Robèrt, Herman Daly, Dana Meadows, and Amory Lovins. Many of their ideas have been incorporated into my own, often inseparably. I don't know where theirs end and mine begin. To that list I have to add Rachel Carson for obvious reasons. I count all the others as personal friends; I only wish I could have known her. The bibliography contains all the works of this group which I have read, along with other works from which I have drawn inspiration.

I think, too, of people who have never written anything for others to read, but who have lived their lives in humble acknowledgment that they were part of nature, and visitors at that. Often ridiculed by more sophisticated neighbors, they nevertheless were right, while their neighbors, people more like I, went blindly and wrongly along trampling Mother Earth underfoot with scarcely a thought.

To environmentalists everywhere, from the giants of the environmental movement to the ordinary folks who *got it* long before I did, I dedicate this little book. Thank you all for the inspiration. Would that I could, in the time I have left as a visitor to this beautiful blue planet, the third from a star called "Sol" in a galaxy called "Milky Way," do justice to your lives and examples with my own.

R.C.A.
July 1998

PROLOGUE

Years ago, there was a popular television show entitled *I Led Three Lives*. It was about the adventures of a businessman and double agent who lived his three lives simultaneously. Well, I, too, have lived three lives, but I have lived them sequentially; and am still in my third. (I mean "lives" from a professional standpoint.)

My first life spanned the first 38 years of my sojourn on Earth, during which I prepared myself, in countless ways, to be the entrepreneur who would found the company that came to be called Interface, Inc.

My second life began with that act of creation—founding a successful company is as creative an act as there is, including the requisite 99 percent perspiration—and went for another 21 years. During that time, Interface, Inc., founded to produce and sell carpet tiles for American office buildings, survived start-up and prospered beyond anyone's dreams. But twice in its 25 years of existence, it has hit the wall; once in 1984 when its primary market—new office construction in the United States—collapsed. That time, the company reinvented and re-formed itself quickly by diversifying its marketplace to include office building renovations and other market segments such as healthcare facilities and other geographical markets, and by entering other businesses, specifically textiles and chemicals, mostly by acquisitions that were financed through public offerings of equity and debt.

Interface hit the wall again during 1991—1993 because of several more or less coincident occurrences: the worldwide movement to downsize corporations, which dampened a key market segment for the company (companies tend not to buy carpet when they are laying off people), worldwide recession, new and tougher competition, and market preferences shifting toward less expensive products that we were not very good at making. The company, once again, reinvented and re-formed itself, but this time with the infusion of a new management team, led by Charlie Eitel, that brought new ideas and new energy. Thus, this second reinvention and re-forming was more profound than the first, and completely changed my life as Founder, Chairman, and CEO of the billion-dollar company that Interface has become.

The new management team took hold of operations quickly and effectively, and my job became one of turning loose, getting out of the way, staying out of the way, and being head cheerleader. That's a big change after 21 years of "nose to the grindstone," autocratic, hands-on management. I began seriously to question my role, what it should be, and if indeed I had one.

Then, in a sequence of events which I describe in the pages that follow, I discovered an urgent calling and an unexpectedly rewarding new role for myself. Thus began my third life, with a new vision of what I wanted Interface, my child, to grow up to be.

One aspect of my new role is being spokesperson to a growing audience that is hungry to hear the Interface story. Trying to satisfy that hunger, I make a lot of public speeches. This little book expresses the beliefs I have embraced and the convictions I have formed as I have prepared those many speeches, all about this urgent calling I refer to as my third life and the vision it has produced for my company and, I fervently hope, will foster for others as well.

CONTENTS

CHAPTER ONE

THE NEXT INDUSTRIAL REVOLUTION

I LIKE TO OPEN my speeches, especially those that deal with the environment, by doing something I learned from my friend, the outstanding family therapist and popular speaker, Dr. J. Zink. I like to have the audience stand and have each person pick out a couple of people standing nearby, preferably strangers, and give them a hug. It never fails to create a lighthearted bond between the audience and me, and it gives me the basis for making a strong opening statement: "I hope the symbolism of that was not lost on you, fellow astronauts on Spaceship Earth. We have only one spaceship. It's in trouble. We're in this together and need each other." According to Dr. Zink, the hug also opens up access to the right hemispheres of people's brains where he says feelings, including conscience, reside. Then, I like to say that, despite the initial levity, my assignment [self-proclaimed] is to disturb, not amuse; to inform, not entertain; and to sensitize (or further sensitize) my audience to the crisis of our times and of all time to come. I invite my audience, if they find me radical and provocative, to be provoked to radical new thinking, and I suggest that all of us need to do more of that.

On a Thursday in April 1996, I was in Boston on a panel speaking to 500 people. The subject was "Planning for Tomorrow," and the panel was about technology's role and impact on the strategic decisions companies make. The discussion was sponsored by the International Interior Designers Association. The audience was about one-third interior designers and two-thirds business people, including some of my company's competitors.

While the subject of the discussion was technology, I think that the audience's understanding of the term probably had to do with the technology in the offices where most of them worked— information technology: office automation, computers, e-mail, radio mail, laptops, word processors, CADs, telephones, voice mail, video conferencing, faxes, Internet, intranets, websites, and so on. There is an infinite variety of gadgets and networks and servers that helps us store information, manipulate information, do arithmetic faster, retrieve information, transmit information, receive information, and examine information—in written form, spoken form, picture form, virtual reality form. Technology gives us faster, surer information when we want it and where we want it, in whatever form we want it. Understanding the information and using it wisely, of course, is then up to you and me. Technology does not do that for us. We're on our own in developing the wisdom and knowledge and understanding to make the information useful.

That's my mental map of what most people, especially people who work in offices, are thinking and meaning when they talk about technology. But I checked out the definition of "technology" in *The American College Dictionary*, and here it is:

1A. THE APPLICATION OF SCIENCE, ESPECIALLY TO INDUSTRIAL OR COMMERCIAL OBJECTIVES.

1B. THE ENTIRE BODY OF METHODS AND MATERIALS USED TO ACHIEVE SUCH INDUSTRIAL OR COMMERCIAL OBJECTIVES.

2. THE BODY OF KNOWLEDGE AVAILABLE TO A CIVILIZATION THAT IS OF USE IN FASHIONING IMPLEMENTS, PRACTICING MANUAL ARTS AND SKILLS, AND *EXTRACTING* [emphasis added] OR COLLECTING MATERIALS.

So, there's quite a lot there that we don't find if we just look in the office: technology that's not electronic, that is not about storing, manipulating, sending, receiving, and examining information. There's chemical technology, mechanical technology, electrical, civil, aeronautical and space technologies, construction, metallurgical, textile, nuclear, agricultural, automotive technologies, now even biotechnology, and so forth.

I illustrated the point for my Boston audience with an example: I told them that I run a manufacturing company that produced and sold $802 million worth of carpets, textiles, chemicals, and architectural flooring in 1995—and would likely sell $1 billion worth in 1996—for commercial and institutional interiors. Now we have offices, too, chock full of office technology:

mainframes, PCs, networks; you name it. And people who are hotelling and teaming, working anywhere, anytime. Information technology makes it all possible, hooking us up around the world.

But we also operate factories that process raw materials into finished, manufactured products that, happily, many members of my Boston audience routinely use and specify for others to use, and our raw material suppliers operate factories. And when we first examined the entire supply chain comprehensively, we found that in 1995 the technologies of our factories and our suppliers', together, extracted from the earth and processed 1.224 billion pounds of material so we could produce those $802 million worth of products—1.224 billion pounds of materials from Earth's stored natural capital. I asked for that calculation and when the answer came back, I was staggered; I don't know how it struck them in Boston, or how it strikes you reading this, but it made me want to throw up.

Of the roughly 1.2 billion pounds, I learned that about 400 million pounds was relatively abundant inorganic materials, mostly mined from the Earth's lithosphere (its crust), and 800 million pounds was petro-based, coming from either oil, coal, or natural gas. Now here's the thing that gagged me the most: roughly two-thirds of that 800 million pounds of irreplaceable, non-renewable, exhaustible, precious natural resource was *burned up*—two-thirds! —to produce the energy to convert the other one-third, along with the 400 million pounds of inorganic material, into products—those products that my Boston friends, and others like them around the world, had specified for others to use or used in their own offices, hospitals, schools, airports, and other facilities. That fossil fuel, with its complex, precious, organic molecular structure, is gone forever— changed into carbon dioxide and other substances, many toxic, that were produced in the burning of it. These, of course, were dumped into the atmosphere to accumulate, and to contribute to global warming, to melting polar ice caps, and someday in the not too distant future to flooding coastal plains, such as much of Florida and, in the longer term, maybe even the streets of Boston (and New York, London, New Orleans, and other coastal cities). Meanwhile, we breathe what we burn to make our products and our livings.

Don't get me wrong. I let that Boston audience know that I appreciated their business! And that my company was committed to producing the best possible products to meet their specifications as efficiently as possible. BUT REALLY, THIS CANNOT GO ON INDEFINITELY, CAN IT? Does anyone rationally think it can? My company's technologies and those of every other company I know of anywhere, in their present forms, are plundering the earth. This cannot go on and on and on.

However, is anyone accusing me? No! No one but me. I stand convicted by me, myself, alone, and not by anyone else, as a plunderer of the earth. But no, *not* by our civilization's definition; by our civilization's definition, I am a captain of industry. In the eyes of many people, I'm a kind of modern day hero, an entrepreneur who founded a company that provides over 7,000 people with jobs that support them, many of their spouses, and more than 12,000 children—altogether some 25,000 people. Those people depend on those factories that consumed those materials! Anyway, hasn't

Interface paid fair market prices for every pound of material it has bought and processed? Doesn't the market govern?

Yes, but does the market's *price* cover the *cost*? Well, let's see. Who has paid for the military power that has been projected into the Middle East to protect the oil at its source? Why, you have, in your taxes. Thank you very much. And who is paying for the damage done by storms, tornadoes, and hurricanes that result from global warming? Why you are, of course, in your insurance premiums. Thank you again. And who will pay for the losses in Florida and the cost of the flooded, abandoned streets of Boston, New York, New Orleans, and London someday in the distant future? Future generations, your progeny, that's who. (Bill McDonough, Dean of the School of Architecture at the University of Virginia, and a leading proponent of "green" architectural design for many years, calls this intergenerational tyranny, the worst form of remote tyranny, a kind of taxation without representation across the generations, levied by us on those yet unborn.) And who pays for the diseases caused by the toxic emissions all around us? Guess! Do you see how the revered market system of the first industrial revolution allows companies like mine to shift those costs to others, to *externalize* those costs, even to future generations?

In other words, the market, in its pricing of exchange value without regard to cost or use value, is, at the very least, opportunistic and permissive, if not dishonest. It will allow the externalization of any cost that an unwary, uncaring, or gullible public will permit to be externalized—*caveat emptor* in a perverse kind of way. My God! Am I a thief, too?

Yes! By the definition that I believe will come into use during the *next industrial revolution.* (I didn't originate that term. Business writer Paul Hawken and architect Bill McDonough have called for "the next industrial revolution," an idea that, as you can see, I have latched onto, because I agree with them that the first one is just not working out very well, even though I am as great a beneficiary of it as most anyone.)

To my mind, and I think many agree, Rachel Carson, with her landmark book, *Silent Spring,* started the *next* industrial revolution in 1962, by beginning the process of revealing that the first industrial revolution was ethically and intellectually heading for bankruptcy. Her exposure of the dangers of pesticides began to peel the onion to reveal the abuses of the modern industrial system.

So, by my own definition, I am a plunderer of the earth and a thief—today, a *legal* thief. The perverse tax laws, by failing to correct the errant market to internalize those externalities such as the costs of global warming and pollution, are my accomplices in crime. I am part of the endemic process that is going on at a frightening, accelerating rate worldwide to rob our children and their children, and theirs, and theirs, of their futures.

There is not an industrial company on earth, and—I feel pretty safe in saying—not a company or institution of any kind (not even, as I told my Boston audience, an interior design firm) that is sustainable, in the sense of meeting its current needs without, in some measure, depriving future generations of the means of meeting their needs. When Earth runs out of finite, exhaustible resources and ecosystems collapse, our descendants will be left holding the empty bag. Someday, people like me may be put in jail. But maybe, just maybe, the changes that accompany the next industrial revolution can keep my kind out of jail. I hope so, most assuredly.

As maybe you can tell, I've seen the light on this—a little late, admittedly. But I have challenged the people of Interface to make our company the first industrial company in the whole world to attain environmental sustainability, and then to become restorative. To me, to be restorative means to put back more than we take, and to do good to Earth, not just no harm. The way to become restorative, we think, is first to become sustainable ourselves and then to help or influence others toward sustainability. Later, I'll show you our map for getting there, The Interface Model for a Sustainable Enterprise.

As for "Planning for Tomorrow"—the theme of that Boston discussion—when we think of the technologies of the future, sustainability, this issue of absolute, overriding importance for humankind, will depend on and require what I believe are the *really and truly* vital technologies, whether developed by us, by our suppliers, or by others like us, the technologies of the next industrial revolution. I don't believe we can go back to pre-industrial days; we must go on to a better industrial revolution than the last one, and get it right this time.

But what does that mean? I have read Lester Thurow's view that we are already in the *third* industrial revolution. He holds that the first was steam powered; the second, electricity powered; making possible the third, which is the information revolution, ushering in the information age. Clearly, all three stages have emerged with vastly different characteristics, and it can be argued that each was revolutionary in scope.

However, I take the view that they all share some fundamental characteristics that lump them together with an overarching, common theme. They were and remain an unsustainable phase in civilization's development. For example, someone still has to manufacture your 10 pound laptop computer, that icon of the information age. On an "all-in" basis, counting everything processed and distilled into those 10 pounds, it weighs as much as 40,000 pounds, and its manufacturers, going all the way back to the mines (for materials) and wellheads (for energy), created huge abuse to Earth through extractive and polluting processes to make it. Not much has changed over the years except the sophistication of the finished product. So I refer to all three of those stages collectively as the first industrial revolution, and I am calling for the next *truly revolutionary* industrial revolution. This time, to get it right, we must be certain it attains sustainability. We may not, as a species, have another chance. Time is short, as we shall see when we get to a discussion of geologic time.

At Interface, we have undertaken a quest, first to become sustainable and then to become restorative. And we know, broadly, what it means for us. It's daunting. It's a mountain to climb that's higher than Everest. It means creating and adopting the technologies of the future—kinder, gentler technologies that emulate nature. That's where I think we will find the model.

Someone has said, "A computer, now that's mundane, but

a tree, *that's* technology!" A tree operates on solar energy and lifts water in ways that seem to defy the laws of physics. When we understand how a whole forest works, and apply its myriad symbiotic relationships analogously to the design of industrial systems, we'll be on the right track. That right track will lead us to technologies that will enable us, for example, to operate our factories on solar energy. A halfway house for us may be fuel cell or gas turbine technologies. But ultimately, I believe we have to learn to operate off current income the way a forest does, and, for that matter, the way we do in our businesses and households, not off of capital—stored natural capital—but off current energy income. Solar energy is current energy income, arriving daily at the speed of light and in inexhaustible abundance from that marvelous fusion reactor just eight minutes away.

Those technologies of the future will enable us to feed our factories with recycled raw materials—closed loop, recycled raw materials that come from harvesting the billions of square yards of carpets and textiles that have already been made—nylon face pile recycled into new nylon yarn to be made into new carpet; backing material recycled into new backing material for new carpet; and, in our textile business, polyester fabrics recycled into polyester fiber, to be made into new fabrics, closing the loop—using those precious organic molecules over and over in cyclical fashion, rather than sending them to landfills, or incinerating them, or downcycling them into lower value forms by the linear processes of the first industrial revolution. Linear must go; cyclical must replace it. Cyclical is nature's way.

In nature, there is no waste; one organism's waste is another's food. Bill McDonough has been saying it for years: "Waste equals food." For our industrial process, so dependent on petrochemical, man-made raw materials, this means "technical food" to be reincarnated by recycling into the product's next life cycle, and the next. Of course, the recycling operations will have to be driven by renewable energy, too. Otherwise, we will consume more fossil fuel for the energy to recycle than we will save in virgin petrochemical raw materials by recycling in the first place. We want a gain, not a net loss.

But if we get it right during the next industrial revolution, we will never have to take another drop of oil from the Earth for our products or industrial processes. That epitomizes my vision for Interface.

Those technologies of the future will enable us to send zero waste and scrap to the landfill. We're already well down this track at Interface. We have become disciplined and focused in all of the businesses that comprise Interface on what is sometimes called the "low-hanging fruit," the easier savings to realize. We named this effort QUEST, an acronym for Quality Utilizing Employees' Suggestions and Teamwork. In the first three-and-a-half years of this effort, we reduced total waste in our worldwide business by 40 percent, which saved $67 million (hard dollars), and those savings are paying the bills for all the rest of this revolution in our company. We are on our way to saving $80 million or more *per year* when we reach our goals.

We're redesigning our products for greater resource efficiency, too; for example, we are producing carpets with lighter face weights (less pile) and *better* durability. It sounds paradoxical, but it's actually working, in a measurable way. We're making carpets with lower pile heights and higher densities, utilizing carpet face constructions that wear better in high traffic, but use less materials— a tiny, but important, step in "de-materializing" business and industry, an intriguing aspect of the next industrial revolution. The embodied energy *not* used in the nylon *not* consumed is enough to power the entire factory making the redesigned products—twice!

Those technologies of the future will enable us to operate without emitting anything into the air or water that hurts the ecosystem. We're just beginning to understand how incredibly difficult this will be, because the materials coming into our factories from our suppliers are replete with substances that never should have been taken from Earth's crust in the first place—as we shall see. But just imagine factories with no outlet pipes for effluent and no smokestacks because they don't need them! Don't you like that? Paul Hawken and Bill McDonough were the first people I heard articulate this concept, too.

Those technologies of the future must enable us to get our people and products from Point A to Point B in resource-efficient fashion. In our company alone, at any hour of the day, we have more than 1,000 people on the move, while trucks and ships (and sometimes planes) deliver our products all over the world. Part of the solution will be Rocky Mountain Institute physicist Amory Lovins' hypercar. When Amory's super lightweight, super aerodynamic hypercar is using solar energy for electrolysis of water to extract hydrogen to power its fuel cells and a flywheel, magnetically levitating at 100,000 rpm, in lieu of a battery, or an ultracapacitor with nothing moving and nothing to wear out, to store energy, including recapturing the energy generated in braking the car, with this energy going to power electric motors on each wheel without any drive train to waste energy, we'll be getting there with an important technology of the next industrial revolution.

To complement and reinforce these new technologies, we will continue to sensitize and engage all 7,000 of our people in a common purpose, right down to the factory floor and right out there face to face with our customers, to do the thousands and thousands of little things—the environmentally sensitive things, energy saved here, pollution avoided there—that collectively are just as important as the five big things, those technologies of the future: solar energy, closed loop recycling, zero waste, harmless emissions, and resource-efficient transportation.

Finally, I believe we must redesign commerce in the next industrial revolution, and redesign our role as manufacturers and suppliers of products and services. Already, we are acquiring or forming alliances with the dealers and contractors that install and maintain our products, requiring an investment of some $100 million in the United States alone since 1995. With these moves downstream into distribution, we are preparing to provide cyclical, "cradle-to-cradle" (another term I borrow from Bill McDonough) service to our customers, to be involved with them beyond the life of our products, into the next product reincarnation, and the next. The distribution system will, through reverse logistics, become, as well, a collection and recycling system, keeping those precious molecules moving through successive product life cycles.

In our reinvented commercial system, carpet need not be bought or sold at all. Leasing carpet, rather than selling it, and being responsible for it cradle-to-cradle, is the future and the better way. Toward this end, we've created and offered to the market the Evergreen Lease®, the first ever perpetual lease for carpet. We sell the *services* of the carpet: color, design, texture, warmth, acoustics, comfort under foot, cleanliness, and improved indoor air quality, but not the carpet itself. The customer pays by the month for these services. In this way we make carpet into what Michael Braungart, a German chemist and associate of Bill McDonough in McDonough Braungart Design Chemistry, terms a "product of service," what Paul Hawken described as "licensing" in *The Ecology of Commerce,* and what the President's Council on Sustainable Development calls "extended product responsibility." Walter Stahel, Swiss engineer and economist, was perhaps the first person to conceptualize such a notion. (There's more about the Evergreen Lease in Chapter Two.)

Environmental sustainability, redefined for our purpose as taking nothing from the earth that is not renewable and doing no harm to the biosphere, is ambitious; it is a mountain to climb, but we've begun the climb. Each of the seven broad initiatives we've undertaken—the five areas of new technologies, sensitized people, and reinvented commerce—is a face of that mountain. Teams all through our company in manufacturing locations on four continents are working together on hundreds of projects and technologies that are taking us up those seven faces toward sustainability, toward the summit of that mountain that is higher than Everest. We know we are on just the lowest slopes, but we believe we have found the direction that leads upward.

We've embraced The Natural Step, the frame of reference conceived by Dr. Karl-Henrik Robèrt of Sweden to define the system conditions of ecological sustainability, as a compass to guide our people up the mountain. In the thousands and thousands of little things, The Natural Step is helping provide what we have termed the "sensitivity hook-up" among our people, our communities, our customers, and our suppliers. We want to sensitize all our constituencies to Earth's needs and to what sustainability truly means to all of us. We want to engage all of them in the climb.

We started this whole effort in our company on two fronts: the first was focused on waste reduction. That's the revolution we call QUEST. It's our total quality management program, and more; the emphasis is broad. We define waste as *any* cost that goes into our product that does not produce value for our customers. Value, of course, embraces product quality, and more—aesthetics, utility, durability, resource efficiency. Since in pursuit of maximum value any waste is bad, we're measuring progress against a zero-based waste goal. A revolutionary notion itself, our definition of waste includes not just off-quality and scrap (the traditional notion of waste); it also means anything else we don't do right the first time— a misdirected shipment, a mispriced invoice, a bad debt, and so forth. In QUEST there is no such thing as "standard" waste or "allowable" off-quality. QUEST is measured in hard dollars and, as

I said, we've taken 40 percent, or $67 million, out of our costs in three-and-a-half years, on our way to a rate of more than $40 million *per year* of waste reduction by the end of 1998, and that much or more again when we actually get to zero waste. One quick result: scrap to the landfills from our factories is down over 60 percent since the beginning of QUEST in 1995, in some factories, 80%.

We've also begun to realize that conceptually it might even be possible to take waste, by its current definition, *below* zero as measured against our 1994 benchmark. If we substitute one form of energy (solar) for another (fossil), or one form of material (recycled) for another (virgin), we are making systemic changes that create, in effect, negative waste when measured against the old norms. If successful, we will have replaced the old system, now obsolete and shown in comparison to have been wasteful all along, with the new, non-wasteful system. So, to give this new meaning to everyday activities, we have further changed our definition of waste in one category and declared *all* energy that is derived from fossil fuels to be waste, waste to be eliminated systematically, first through efficiency improvement and, eventually, to be replaced by renewable energy. Even the irreducible minimum of energy needed to drive our processes is waste by this definition, as long as it comes from non-renewable sources. QUEST *is* a revolution in operational philosophy.

The second parallel effort we've called EcoSense®. It's focused on those other four major technologies of the future, together with the thousands of little things and the redesign of commerce. Measurement is more difficult for EcoSense. We're dealing here with "God's currency," not dollars, guilders, or pounds sterling—the field called EcoMetrics®, a term I coined. Here's an example of EcoMetrics: How do you evaluate the following hypothetical trade-off? One product consumes 10 pounds (per unit) of petrochemically derived material, a non-renewable resource. Another, functionally and aesthetically identical to the first, consumes only six pounds, substituting four pounds of abundant, benign, inorganic material, but through the addition of a chlorinated paraffin. That chlorine could be the precursor of a deadly dioxin. How does one judge the true cost or value (which is it?) of that *chlorinated* paraffin—in God's currency? That's EcoMetrics, the search for God's currency. It's perplexing—a scale that weighs such diverse factors as toxic waste, dioxin potential, aquifer depletion, carbon dioxide emissions, habitat destruction, non-renewable resource depletion, and embodied energy. EcoMetrics: we need God's own yardstick, and wisdom, to help us measure where we are, which direction we're headed, and to tell us when we reach sustainability. Dollars and cents alone won't tell us.

In February 1996, we brought these two revolutionary efforts, QUEST (the hard dollar effort) and EcoSense (the "God's currency" effort, measured by EcoMetrics) together. We merged the two task forces into one, and formed 18 teams with representatives from all of our businesses worldwide, each team with an assigned scope of investigation. It was a wonderful marriage. It is integrating these closely related efforts and positively changing our corporate culture because it is making us think differently about who we are and what we do. As my associate, Dr. Mike Bertolucci, says, "It is as if you enter every room through a different door from the usual one, so different is the perspective from which you view every opportunity." I call it "piercing the veil" and finding on the other side a whole new world of opportunity and challenges. Today there are more than 400 projects, from persuading our landlord to install compact florescent light bulbs in our corporate headquarters office, to creating new, sustainable businesses within our company.

Other companies, different from ours, will have to pursue different technologies, different from ours. In the next industrial revolution, I believe they must if they expect to survive. In the 21st Century, as the revolution gathers speed, I believe the winners will be the resource-efficient. At whose expense will they win? At the expense of the resource-*in*efficient. Technology at its best, emulating nature, will eliminate the inefficient adapters.

Meanwhile, the argument goes on between technophiles and technophobes, one saying technology will save us, the other saying technology is the enemy. I believe the next industrial revolution will reconcile these opposing points of view, because there is another way to express the differences between the first industrial revolution and the next. The well known environmental impact equation, popularized by Paul and Anne Ehrlich in their writings, declares that:

$$I = P \, x \, A \, x \, T$$

In the equation, I is environmental impact (bigger is worse), P is population, A is affluence, and T is technology. An increase in P, A, or T results in a greater (worse) environmental impact. Technology *is* part of the problem, the technophobes' position. But that is the technology of the first industrial revolution, call it T_1. Now the equation reads:

$$I = P \; x \; A \; x \; T_1$$

What a dilemma! T_1 is not the answer. T_1 will not lead us out of the environmental mess, no matter how vigorously the technophiles assert it will. The more technology we have, the greater (worse) the impact. Remember that "10 pound" laptop computer and the extractive, abusive processes that produced it?

But just what are the characteristics of T_1, the technologies of the first industrial revolution? For the most part, they are extractive (written right into the dictionary definition of technology), linear (take-make-waste), fossil fuel-driven, focused on labor productivity (more production per worker), abusive, and wasteful—the destructive, voracious, consuming technologies of the first industrial revolution. And they are unsustainable.

But what if the characteristics of T were changed? Call it T_2 now, the technologies of the next industrial revolution. Let's say they were *renewable*, rather than extractive; *cyclical* (cradle-to-cradle), rather than linear; *solar- or hydrogen-driven*, rather than fossil fuel-driven; focused on *resource productivity*, rather than labor productivity; and *benign* in their effects on the biosphere, rather than abusive. And what if they *emulated nature*, where there is no waste?

Mightn't it then be possible to restate the environmental impact equation as:

$$I = \frac{P \; x \; A}{T_2}$$

Wow! Then the technophiles, the technophobes, the industrialists, and the environmentalists could be aligned and allied in their efforts to reinvent industry and civilization. Move T from the numerator to the denominator and we change the world as we have known it. The mathematically minded see it immediately. Now, the more technology the better (less impact). Furthermore, it begins to put the billion unemployed people of Earth to work—working on increasing resource productivity, using an abundant resource, labor, to conserve diminishing natural resources. Technology becomes the friend of labor, not its enemy. Technology becomes part of the solution rather than part of the problem. Again, I credit Bill McDonough for a magnificent insight: T must move to the denominator.

What will drive technology (T) from the numerator to the denominator? I believe getting the prices right is the biggest part of the answer; that means tax shifts and, perhaps, new financial instruments such as tradable emission credits, to make pollution cost the polluter—in effect, a carbon tax. In any event, it means eliminating the perverse incentives and getting the incentives right for innovation, correcting and redressing the market's fundamental dishonesty in externalizing societal costs, and harnessing honest, free market forces. If we can get the incentives right, entrepreneurs everywhere will thank Rachel Carson for starting it all. There are new fortunes to be made in the next industrial revolution.

But what in turn will drive the creation of tax shifts and other politically derived financial instruments? It seems to me that those will ultimately be driven by a public with a high sense of ethics, morality, a deep-seated love of Earth, and a longing for harmony with nature. When the marketplace, the people, show their appreciation for these qualities and vote with their pocketbooks for the early adopters, the people will be leading; the "good guys" will be winning in the marketplace and the polling booth; the rest of the political and business leaders will have to follow. As a politician once said, "Show me a parade and I'll gladly get in front of it." So will business and industry respond to the demands of this new marketplace, and Earth will gain a reprieve.

CHAPTER TWO

A SPEAR IN THE CHEST AND SUBSEQUENT EVENTS

FROM MY PERSPECTIVE—presumptuous, perhaps—this journey, this climb, has become an epic story. I've been told by Dr. Zink that all epic stories begin *in medias res*. So, think of what I've just said in Chapter One as the middle of things for my company, Interface, and me. What about the beginning? How did we get "here" from "there"? What was "there"? What is "here"?

Please indulge me as I switch to a personal vein. Someone has said that everybody has just one story to tell, her or his own story. Here is part of my story.

I was born and grew up in the small, west Georgia town of West Point, the third of three sons—"Baby Ray." I am also a product of the Great Depression, the era into which I was born, and World War II, as well as the post-war era which was, of course, one of enormous prosperity and economic opportunity in America.

However, that was still over the horizon in the mid 1930s, and times were tough. My father, William Henry Anderson, was the oldest of seven children and had been made, as was often expected of the eldest son in those days (the early decades of the 20th Century), to sacrifice his own education and go to work to help his father support the family. So, after the eighth grade, my father had to quit school and go to work to enable his sisters eventually to go on to college. Even as a child, that struck me as a grotesque waste. Though I never heard him complain, I believe he knew it was a waste. That awareness shaped his determination not to let his sons waste their lives. He made sure that my brothers and I got an education. One brother, Bill, Jr., became a medical doctor, the other, Wiley, a teacher. I became an industrialist.

My mother, Ruth McGinty Anderson, was also one of seven children, the middle child of more enlightened parents. Her eldest brother excelled in the pulpit and other brothers excelled in business. She became a school teacher, but (again as was customary in those days) was not allowed to teach school after she married. So, she practiced her profession on her three sons. I responded well and loved school.

I grew up with a book in one hand and a ball in the other. Whatever ball was in season—football, basketball, baseball, and softball, in succession—occupied my days, and books occupied my evenings. It was the football that ultimately assured my opportunity to go to college. I earned a football scholarship to Georgia Tech, being good enough as a running back to capture the eye of legendary coach Bobby Dodd during his golden years at Georgia Tech.

However, it was my friendship with books, together with the study discipline imparted by my mother and my teachers in the public schools of West Point, that transformed that scholarship opportunity into an excellent education.

Ironically, it was an eighth grade experience with football, in 1947, that was a defining moment for me—at the same age my father had been when his life was essentially defined for him by his father. I was big for my 13 years, weighing in at 142 pounds, and the high school coach, Carlton Lewis, urged me to go out for football to develop my skills and to gain experience practicing against the really big boys, the high school varsity team. So, I went out and very soon found myself scrimmaging against those guys— the future state champion team—grist for the mill, so to speak. Playing defensive linebacker one day, I encountered the biggest, hardest-charging of the varsity running backs, one who went on to become an all-state and college level player, at the sideline—my head against his knee in a thunderous collision. The collision hurt us both, his knee and my head. I could look up and see the knot rising on my forehead with each beat of my heart. That ended practice for both of us that day.

As far as I was concerned, it also ended football for me forever. The next day I did not go to practice, but Coach Lewis would have none of that. He left the practice field, found a telephone in a nearby house, and called my father at work in the post office. My father, justifiably glad to have obtained secure employment (dating from depression days when secure jobs were hard to find) and having risen to Assistant Postmaster, never ever had left his job for anything.

But he did that day, and he walked the streets of our small town (we didn't own a car) until he found me. He shamed me completely for quitting, with the tongue-lashing of my life, and the next day I was back at football practice, with sore forehead and chastened spirit, duty-bound not to waste my life with a bad eighth grade choice. I never liked football again, but I played hard and successfully until a shoulder injury my sophomore year at Tech ended my playing career. I laid down my last ball and, with the hand thus freed, took up preparing myself for business, holding on to that book in the other.

I loved my mother and my teachers; I loved but resented my father for making me do something I hated. In my heart I felt that he, already financially burdened with two sons' college educations, saw my budding athletic ability as the means for shedding the third such burden. I hated my high school coach, but grew to respect (and perhaps love) him in time for making me better than I wanted to be, and—in a truly defining way—teaching me, as he taught everyone who came under his influence on the athletic field, to compete.

No other lesson in my entire life has been more valuable than that one. It is no accident that a hallmark of Interface today is the never-say-die, never-ever-give-up attitude of its sales force in pursuit of the next "heartbeat" for our company, the next order, and the goodwill of the loyal customer that provides that heartbeat, again and again.

It was competition at the academic level, beginning in the fourth grade, that further prepared me to excel at Georgia Tech. A new child joined our tiny class that year, and from her first day, Barbara Adams knew more than I did—always just a little bit more. Eight years of intense, head-to-head struggle later, Barbara nosed me out by 0.1 on a scale of 100.0 for valedictorian honors in our graduation exercise. That competition on the book side of things was just as valuable a lesson as Carlton Lewis' unrelenting pressure on the athletic field. Today, Barbara Adams Mowat is Head of Academic Programs at the Folger Shakespeare Library in Washington, D.C., Senior Editor of *Shakespeare Quarterly*, Co-Editor of *The New Folger Library Shakespeare*, and America's foremost authority on Shakespeare, one of an elite handful in the entire world. I like to think I had something to do with that—as she influenced me—stemming from that struggle between two friends to be number one.

I worked hard at Georgia Tech, made Tau Beta Pi engineering honorary society, and graduated in 1956 with Highest Honors and a bachelor's degree in Industrial Engineering. I then spent the next 17 years climbing the corporate ladder and preparing myself by gaining broad business experience, but mostly subconsciously, to take the plunge and become an entrepreneur. In 1973, I was 38 years old and had a very good job with a major corporation. I left that job and company and cut the corporate umbilical cord to found a new company to produce, of all things, free-lay carpet tiles. The act of cutting that cord required one of the two hardest decisions of my life, a decision that had developed over the course of nearly seven years.

I had done well quickly in my climb up the corporate ladder. Four years after graduation from Tech, after unexciting jobs with two other companies, I found myself on a fast track upward, thanks to being picked out of a crowd of "new blood" to which I had been recruited as a part of the rejuvenation of Callaway Mills Company, a Georgia-based textile manufacturer. In 1959, Fuller Callaway, Jr., Chairman and CEO, began the process, though few of us recognized it, of preparing his family foundation-owned company to be sold. Nine years later the sale occurred, but in the meantime it was "Camelot" for the management team that Mr. Callaway had hand picked to lead the process of pumping up the company for sale. But Camelot came to an abrupt end for me when I got passed over in favor of another executive for the job I most wanted: to head one of the three operating divisions of Callaway Mills. I was stung! To put it bluntly, being passed over chapped my ass!

That was in 1966, and my job as Vice President, Staff Manager—responsible for all the non-financial staff functions of a company with $80 million in sales—no longer was enough to satisfy my ambitions. Wounded by the decision that disrupted my climb and over which I felt I had had no control, I began the psychological journey toward entrepreneurship—to do my own thing and make my own decisions, especially those affecting my own career destiny. Two years later, on April 1, 1968 (that's right, April Fools' Day), Callaway Mills was acquired by a much larger textile company, Deering Milliken. All of my staff functions were quickly absorbed into their counterpart departments at Milliken, and soon I was reassigned to become Director of Development of

Milliken's Floor Covering Business. Within a year I fell in love with a new idea, which I saw for the first time in June 1969 in Kidderminster, England—carpet tiles, a new concept for covering an office floor.

I took a leading role in helping Milliken bring carpet tiles to the United States from Europe. It was a major development project, and by 1972 Milliken was the established leader in the United States in this emerging niche market. After nearly seven years of searching, triggered by that 1966 disappointment, I had found *my* thing, too. Carpet tiles were so right! So smart!

For a year I corresponded with the Kidderminster, England carpet manufacturing company, Carpets International (CI), and finally persuaded them to join me in a venture to bring their patented carpet tile technology to the United States. As I write this, it is just about exactly 25 years ago that Jim Carpenter, CI's Vice-Chairman, called from England and woke me up on a January morning in 1973, to tell me that CI was ready to go with the venture that we had been discussing and planning. Only there were some provisos. I must leave Milliken without legal impediment. I must retain legal counsel and public accountants satisfactory to CI. I must arrange a banking facility for the debt portion of our planned capitalization. And I must raise $500,000 of equity to go with the $750,000 that CI was prepared to invest in the proposed 50/50 Anglo-American venture.

That last one was a tall undertaking, much taller than I first thought. During the latter stages of our planning in late 1972, Smith Lanier and Robert Avary, both friends from my hometown, had been very supportive. With Robert's high level of interest and wealth to back that up, Smith and I thought that raising $500,000 would not be a big problem. So, on January 19, 1973, I invested the first $50,000 out of my life savings (which, at the time, totaled about $65,000) to fund the initial capitalization of Compact Carpets, Inc., the predecessor of the predecessor of Interface.

We thought that my ability to satisfy another of CI's provisos—that I leave Milliken without any legal impediment—might be more difficult. After all, that depended on Milliken as much as on me.

On February 1, 1973, I tendered my two-weeks notice to Milliken, and told them generally what I planned to do, i.e., create a new company from scratch, with CI as a partner, to compete with them in the emerging new product/market niche: free-lay carpet tiles for the "office of the future." For the rest of that week and the next no one spoke to me, until Friday, February 12, my last day, when two of my fellow managers walked into my office, closed the door, and sat down. They told me that they did not believe I could legally do what I was planning to do, that I would unavoidably use proprietary information I had acquired at Milliken in any such new venture, and that such information was covered by a broad secrecy agreement I had entered into as a condition of employment with Milliken (following Milliken's acquisition of Callaway). Milliken had acquired different, competitive carpet tile technology, but the technologies were close enough that their concerns were legitimate.

More defiantly than I felt, I replied, "The hell you say!" And I invited them to bring their lawyer to Atlanta to meet with mine,

Carl Gable, the next week. We set a date—Tuesday, February 16, 1973, in Carl's office.

Over that weekend, Robert Avary dropped out on me as an investor. He had by then committed to be my "angel" and to take any part of the $450,000 balance-to-raise for which I could not line up other investors. But he called me on Saturday from the psychiatric ward of Emory University Hospital to tell me that his wife had filed for divorce, that his financial holdings were "tied up," and that he had checked himself into the hospital. "Your need for money could not have come at a worse time," was his heart-stopping pronouncement.

What to do? What a dilemma! The Milliken meeting coming up, my bridges burned, and now this! What a stomach-knotting turn of events.

That's when Smith stepped up like the champion he was, and still is, and said, "Don't worry, we'll get the money." Those words are branded in my brain for as long as I will live. "Don't worry, we'll get the money," the classic link between entrepreneur and capitalist.

So Carl and I met with the Milliken people Tuesday. Carl challenged their position with, "What do you mean? Are you saying we can't compete?"

They quickly backed off. "No, we didn't mean that." And they tried very hard to entice me back with promises to forgive everything and treat me as if I had never left. It was very tempting. These were turbulent waters. Carl found me a private room when I asked for time to think about it one more time, to "review the bidding," so to speak. It was Smith's admonition, "Don't worry, we'll get the money," that was the cornerstone of my decision in that lonely room to move ahead. It was a pivotal moment. I returned to the meeting room, told the waiting group that I was going ahead, and asked Carl to undertake to find a way to satisfy Milliken's discomfort.

Joe Kyle, whom I had recruited to head up manufacturing for the prospective venture, and I went on to England. We spent the next three weeks finalizing our manufacturing plan with the help of CI technicians and getting a final decision from Peter Anderson (Chairman of CI, and no relation to me) to go forward with the investment. It was a close call for Peter. By the end of the three weeks, Carl Gable had worked out an agreement with Milliken not to oppose the venture legally, provided I agreed not to use or disclose their most sensitive proprietary information, boiled down through long negotiations to their fire retardant pvc formulations. That agreement, together with faith in the abilities of his technical people to work out such details and his own burning desire to get his carpet tiles into the American market—after *three* aborted negotiations with other prospective American partners, including Milliken—persuaded Peter Anderson to take the plunge. I believe Peter took it over the objections of his closest advisors, including Jim Carpenter, his number two man, who had been my first supporter at CI. Undoubtedly, it was a close and gutsy call for Peter; sadly, he did not live to see the astonishing success that flowed from that decision. Interface came oh, so close to *not being* on so many occasions; it's a miracle that it *is*.

However, even with CI aboard, we still had the money issue to deal with—the $450,000 of American equity to be raised. Joe Kyle and I came home in early March, and Smith and I went to work on cultivating prospective investors. Smith and his family led with the first $200,000 commitment. Together with my initial $50,000, which was financing our current expenses, we were halfway to our objective. We found ourselves subject to an SEC rule that limited the number of potential investors we could *even approach* with our private placement offering memorandum to 25. One by one, carefully screening and rationing our 25 contacts jealously, we got 18 people to invest the remaining $250,000.

On April 6, 1973, the full commitment in hand, including a banking arrangement with The First National Bank of Boston (another CI proviso met), the CI money came in and Interface was capitalized under names long since abandoned: Carpets International-Georgia, Inc. and its sister company, Carpets International-Georgia (Sales), Inc., CIGI and CIGI (Sales) for short. CIGI was owned 51/49, CI/Americans, and CIGI (Sales) was owned 51/49, Americans/CI. With the equity evenly divided between the two, we had, effectively, a 50/50 venture with a built in tie-breaker. In a loggerhead disagreement, we could go our separate ways.

My American backers even agreed to go a step further and enter into a Voting Trust Agreement, giving me, in effect, the vote of their 30 percent. Together with the 20 percent I owned, including "sweat equity," that made me an equal partner with CI.

Eight years later, CIGI and CIGI (Sales) would be merged into Interface Flooring Systems, Inc., as the American investors were able to buy 10 of CI's 50 percent to make the venture 60/40, Americans/CI. Five years later we would take over CI completely, get their holdings of our stock back, sell off CI's worldwide operations, and buy Guilford of Maine with the salvaged proceeds. That series of transactions, spanning three years, is a business school case study in its own right.

To tie up a loose end, in due course (with no help whatsoever from me and very little from CI), Joe Kyle worked out our own fire retardant formulations, and we were able to produce competitive products that met all the fire codes. To this day, I have never told another soul what I learned about pvc formulations from Milliken.

Peter Anderson, Smith Lanier, Carl Gable, and Joe Kyle are, without a doubt, on the very short list of people about whom I can say without fear of contradiction: Without them, there would be no Interface.

Still, even with all that support, the risk was so frightening, like stepping off a cliff in the dark and not knowing whether your foot would land on solid ground or thin air. Furthermore, throughout those tense months, beginning in late 1972, it had become increasingly clear that my wife, Sug, did not want me to do this thing I was so intent on doing. In fact, she had become adamantly opposed. We had two children, Mary Anne, 16, and Harriet, 12. College was coming soon. Anyway, didn't I have a good job? Couldn't I just be content with that? Why, why, why?

Good questions. To this day I don't have good answers, other than that I was driven to do it, to seize the opportunity to do my own thing. In the midst of a heated argument with Sug one night, while the whole process was still unsettled, I stormed into an adjacent room and fell to the floor in abject anguish, wrestling with myself: To do it or not to do it? I got up from that floor and shouted aloud to no one but myself, "By God! I'm going to do it!" It was the hardest decision of my life up to then.

On April 6, 1973, when our equity capital was in hand and our banking facility was in place, Interface was born. Everything to that point, beginning with a gleam in my eye, had been conception and gestation. The birth was long, difficult, and very painful and could only be called complete when I had the initial capital in hand.

The decision that surpassed the earlier one to become the hardest of my life came 10 years later when I filed for divorce. The bitter seeds were sown in those arguments in 1972 and 1973 and in the vehement opposition I encountered from my wife at this turning point in my life. From that time on, we grew apart.

Sug had every reason to object. Her protests and sense of insecurity were completely justified, but her opposition served only to spur me on. Not only did I never want to have a boss again to control my career, but I could not let myself fail and face, "I told you so," from her. My self-fulfillment didn't come without a price. The mother of my children paid a price, too, for my shot at success on my own terms. The children themselves, with lives and families of their own (there are now five terrific grandchildren: Jay, John and Patrick, sons of Mary Anne and Jaime Lanier; and Melissa and McCall, daughters of Harriet and Phil Langford), responded with great maturity to the divorce and have accepted Pat, my wife of 14 years now, more as close friend than stepmother.

Born in maelstrom, so close to not being born at all, Interface is precious to me.

❖ ❖ ❖ ❖ ❖

More about this idea with which I fell in love in June 1969: Carpet tiles, modular carpet, came in 18-inch squares (in Europe, 50 cm. squares) that could be installed without adhesive to gain the appearance of broadloom carpet but the easy, flexible functionality of modularity. At that time, carpet tiles were just beginning to be used in American office buildings, where the electrical wiring was in the floor, the furniture was open plan systems furniture, and the office was becoming computerized. The new concept was known, even in those days, as the "office of the future." The office of the future needed carpet tiles for easy access to wiring in the floor and for their everyday practicality. The timing and the product concept were perfect. So right, so smart!

The new venture was an entrepreneur's dream, except for the tension at home: beginning with the idea that carpet tiles were a better mousetrap and the time was right for them in America, followed by not only satisfying those initial provisions and birthing the company, but also by acquiring a site, building and equipping a factory, securing raw materials in a time of extreme scarcity, developing and producing those first American-made products, beginning to build an organization of people—Joe Kyle in Manufacturing, Don Russell in Marketing, Don Lee in Administration—and launching a sales and marketing effort into the teeth of the worst recession since 1929, working like hell—and surviving! I found pioneering a new product in a new market to be the most frightening and stressful, yet exhilarating and highly rewarding experience imaginable. Our major competitor? Who else but Milliken! It was an in-your-face time. Compete, compete, compete! Beat Milliken, beat Milliken! Horatio Alger's stories were no better than this one.

Survive, we did, and we prospered beyond anyone's wildest dreams. Today that company is global. We produce in 29 manufacturing sites, located in the United States, Canada, the United Kingdom, Holland, Australia, Thailand, and (our newest factory) China. We sell our products in more than 110 countries. Sales in 1995 exceeded $800 million; in 1996, they topped $1 billion for the first time; in 1998 they will likely exceed $1.3 billion. We make and sell about 40 percent of all the carpet tiles used on earth and enjoy the largest market share in nearly every one of those 110 countries. After numerous acquisitions over the years, we also produce commercial broadloom carpets, textiles, chemicals, and architectural products (specifically raised access floors) with some of the great brand names in the commercial interiors industry: Interface, Bentley, Guilford of Maine, Prince Street Technologies, C-Tec, Heuga (in Europe), Intek, Toltec, Stevens Linen, and Camborne (in Europe also), and distribute our carpet products through our owned and licensed distribution channel, Re:Source Americas. Most recently, we acquired the Firth and Vebe Operations in Europe from U.K.-based Readicut plc. Our brands are not household words because our products are not used where people live; they are used where people work. We are the world's largest producer of contract commercial carpet. It is an "only in America" success story, another case study yet to be written for aspiring entrepreneurs.

But as successful as it seems, Interface is flawed. It took me a very long time to realize it, though. For the first 21 years of our company's existence, I, for one, never gave one thought to what we were taking from the earth or doing to the earth, except to be sure we were in compliance and keeping ourselves "clean" in a regulatory sense and obeying the law, and to be sure we always had access to enough raw materials, mostly petrochemically derived, to meet our needs. We had very little environmental awareness. Until August of 1994.

True, before that, we had been developing, for fully 10 years, a program called Envirosense®. Envirosense, working through a consortium of companies, had been focused on indoor air quality (IAQ) and alleviating Sick Building Syndrome and Building Related Illness, such as Legionnaire's Disease. This effort had been based on some proprietary chemistry we had acquired in the field of anti-microbials, called Intersept® (yes, with an "s"). Intersept is an additive which, if incorporated into plastic materials, renders the surface of those materials self-sanitizing. So, with Intersept, materials such as carpets, paints, fabrics, air filters, and air duct liners, as well as cooling coil and drip pan coatings can be made to be more hygienic. Better hygiene leads to better air quality by reducing bacterial and fungal growth, contributing to healthier indoor environments by tackling the microbial contamination piece of the very complex IAQ equation. We were accomplishing good things in the field of IAQ. That and compliance were *it* for us, in terms of environmental sensitivity.

But then, in August 1994, some of the people in Interface Research Corporation, our research arm, in response to customers who were beginning to ask what we were doing for the environment—questions for which we did not have adequate answers—decided to organize a task force with representatives from all of our businesses around the world. Our research people wanted to review Interface's company-wide, worldwide environmental position, and begin to frame a response to those customers' questions we could not answer very well. One of these associates, Jim Hartzfeld, suggested that the new task force ask me to make the keynote remarks, to kick off the task force's first meeting and give the group an environmental vision. Well, frankly, I didn't have a vision, except "obey the law, comply, comply, comply," and I was very reluctant to accept the invitation. The idea that, while in compliance, we might be hurting the environment simply hadn't occurred to me. Though I had heard Henry Kissinger advocate as early as 1992 that "sustainable development" should become the galvanizing cause for the West, replacing the Cold War in the new era of peace, I had no idea what he had meant. (I wonder today if *he* knew.) So, I sweated for three weeks over what I would say to that group.

Then, through what seemed to be pure serendipity, somebody sent me a book: Paul Hawken's *The Ecology of Commerce*. I read it, and it changed my life. It hit me right between the eyes. It was an epiphany. I wasn't halfway through it before I had the vision I was looking for, not only for that speech but for my company, *and* a powerful sense of urgency to do something to begin to correct the mistakes of the first industrial revolution. Hawken's message was a spear in my chest that is still there.

Later, I came to realize that it had touched me for another reason. At age 60, I was beginning to look ahead subconsciously to a day that would come soon enough when I would be looking back at the company I would be leaving behind. What would my creation, this third child of mine, be when it grew to maturity? I was looking, without realizing it, for *that* vision, too. A child prodigy in its youth, would it become a virtuoso? What would that mean? These were and are strategically important questions to me, personally, as well as to Interface, Inc.—in the highest sense of the word, strategic. I'm talking about *ultimate purpose.* There is no more strategic issue than that.

In preparing to make that kick-off speech, I went beyond compliance in a heartbeat. I incorporated many examples from *The Ecology of Commerce* to explain what is happening to the ecosystem, using Hawken's description of the reindeer of St. Matthew Island to illustrate such basic concepts as *carrying capacity, overshoot,* and *collapse,* and as an arresting, frightening metaphor for Earth:

A haunting and oft-cited case of . . . an overshoot took place on St. Matthew Island in the Bering Sea in 1944 when 29 reindeer were imported. Specialists had calculated that the island could support 13 to 18 reindeer per square mile, or a total population of between 1,600 and 2,300 animals.

By 1957, the population was 1,350; but by 1963, with no natural controls or predators, the population had exploded to 6,000. The original calculations had been correct; this number vastly exceeded carrying capacity and was soon decimated by disease and starvation. Such a drastic over-shoot, however, did not lead to restabilization at a lower level, with [just] the "extra" reindeer dying off. Instead, the entire habitat was so damaged by the overshoot that the number of reindeer fell drastically below the original carrying capacity, and by 1966 there were only 42 reindeer alive on St. Matthew Island. The difference between ruminants and ourselves is that the resources used by the reindeer were grasses, trees, and shrubs and they eventually return, whereas many of the resources we are exploiting will not.

I cited Hawken's litany of abuse of the earth that we are witnessing in our times:

- The depletion of the Ogallala aquifer, that great underground body of fresh water under the American Midwest, and the implications of that, namely famine right in our own country. All over the world our aquifers are being dehydrated or, worse, polluted.
- The worldwide loss of 25 billion tons of topsoil every year, equivalent to all the wheat fields of Australia disappearing, and a hungry world population increasing by 90 million

a year (now more like 80 million, but still...)

- The usurpation of a disproportionate share of Net Primary Production, the usable product of photosynthesis, by the human species—one species among millions of species taking nearly half for itself—and pushing the ecosystem toward overshoot and collapse for thousands, maybe millions, of species.

- The result: an alarming increase in the rate of species extinction, now between 1,000 and 10,000 times the average rate since the mass extinction of the dinosaurs 65 million years ago. As many as a quarter of all species of plants, animals, and microorganisms on Earth—millions of species—are likely to be lost within a few decades; as many as three-quarters face extinction in the 21st Century. "The Death of Birth," Hawken called it. That phrase brought tears to my eyes when I first read it. It was the very point of the spear. The *death* of *birth*?!! Species lost, never ever again to be born. In no way can this bode well for our own species, because we are fouling our own nest, too.

- The cutting of vast areas of natural forests in Brazil, a critical lobe of Earth's lungs, to clear land to raise soybeans to feed cows in Germany to produce surplus butter and cheese that piles up in warehouses, while a million displaced forest people live in squalor in the *favelas* (ghettos) of Rio de Janeiro. (I cried openly when I read that, and I was astonished and saddened still further to actually see *favelas* on a recent visit to Rio.)

• Illnesses from pesticide poisoning numbering in the millions each year, with uncounted deaths resulting.

In making that first speech, I borrowed Hawken's ideas shamelessly. And I agreed completely with his central thesis: that business and industry, together the largest, wealthiest, most powerful, most pervasive institution on Earth, and the one doing the most damage, must take the lead in directing Earth away from the route it is on toward the abyss of man-made collapse. I gave that task force a kick-off speech that, frankly, surprised me, stunned them, and then galvanized all of us into action. With and through them we are energizing our whole company to step up to our responsibility to lead. Unless somebody leads, nobody will. That's axiomatic. I asked, "Why not us?" Their answer has become a tidal wave of change in our company.

I offered the task force a vision: Interface, the first name in industrial ecology, worldwide, through substance, not words. I gave them a mission: to convert Interface into a restorative enterprise, first to reach sustainability, then to become restorative—putting back more than we ourselves take and doing good to Earth, not just no harm—by helping or influencing others to reach toward sustainability. And I suggested a strategy (you know this one, at least in part): Reduce, reuse, reclaim, recycle (later we added a very important one, *redesign*), adopt best practices, advance and share them. Develop sustainable technologies and invest in them when it makes economic sense. Challenge our suppliers to do the same. And I encouraged them to pick the year by which Interface would achieve sustainability. Two days later they told me their target year, the year 2000. I'll be 66 that year, and would love to see it happen by then. Truthfully, I think they were overly ambitious and that it will take much longer. The enthusiasm of the moment, coupled with a generous measure of naiveté as to the magnitude of the undertaking, led to that initial, optimistic goal. However, I come from long-lived people. The view from the top of that mountain that is higher than Everest will be beautiful beyond words! I hope to live to see it.

We gave this effort the name I've already mentioned, EcoSense. We are taking EcoSense throughout our company, hoping to involve everyone. It's not easy to get 7,000 associates to accept a role in a cause to do the right thing, and I doubt that every single one actually has. I cannot dictate what someone will believe in her or his heart, and that's where every individual decision lies. I just keep urging. For the first year, that urging yielded only barely perceptible effects outside that initial core group, but then the momentum began to gather. Our people, one by one, caught the vision. For the last three years the progress has been phenomenal.

We coined a word, PLETSUS®, an acronym for Practices LEading Toward SUStainability, and we began to share PLETSUS ideas, internally and externally. We'll share them with you. You can go into Interface's website on the Internet and find EcoSense and PLETSUS ideas, right there for you and the rest of the world to see and use. Feel free, and share yours with us. We'd be very happy if PLETSUS and our website caught on and became a worldwide clearing house for idea sharing. Our address is: http://www.ifsia.com. (That's our NASDAQ stock market symbol, too, IFSIA.) A current list of PLETSUS ideas is provided in the Appendix of this book.

Though you can share what we're doing through the Internet, EcoSense is basically our *internally* focused effort to do what's right. But it's not just the right thing to do; it's also the *smart* thing for a manufacturing company that is as dependent as we are on non-renewable resources (petroleum, coal, and natural gas) for its raw materials and its energy-intensive processes. Like carpet tiles in the beginning, EcoSense is so right, so smart. Only now, so much more is at stake—orders of magnitude more. Bill Young, who represents our outside accountants and has been another ally for 25 years, has observed that "survival," an early preoccupation for us, has taken on new meaning, also orders of magnitude more important, as we try now to do our small part for the survival of the species.

❖ ❖ ❖ ❖ ❖

I made other speeches in the months that followed the first one, patterned after that kick-off address, all to Interface people, to begin to bring them aboard. My first outside public speech was to a group of Georgia Tech alumni and faculty; afterward, one of the professors in the audience, Dr. David Clifton, sent me a copy of Daniel Quinn's book, *Ishmael*. I read it, then read it again. I've read it six times, now, and I've bought and given away some 500 copies, always with the admonition, "Pass it on!" I'm here to tell you that Hawken and Quinn, together, will not only change your life, but make you understand why it should change. They did both for me. If you haven't already done so, read *Ishmael* to understand why the world is in this environmental mess. Hawken will tell you *what* the mess is; Quinn, *why*. Those two books should be packaged and sold together, re-titled *The What and the Why of the Biggest Mess of All Times.*

In July of 1995, I met Paul Hawken. What a guy! We've become good friends. He told me that before writing *The Ecology of Commerce,* he read over 200 books and 1,000 papers, altogether more than 25 million words on the environment. He's distilled a lot into that monumental work. It's worth your time to read it if you haven't. Daniel Quinn, too, has become a good friend. His later book, *The Story of B,* was also powerful, the philosophical sequel to *Ishmael,* written in an entirely different, but very compelling, way. Still another one from Quinn is just out: *My Ishmael.* I'm reading it now and loving it. Hawken has another one coming, too, co-authored with Amory Lovins and Hunter Lovins (Amory's wife and associate at Rocky Mountain Institute). It is called *Natural Capitalism,* and it will be an important work.

There was so much to learn that first year! I continued to read, going back to Rachel Carson's seminal *Silent Spring,* Vice President Gore's *Earth in the Balance, Beyond the Limits* by Meadows, Meadows, and Randers (really scary!), *Vital Signs* and *State of the World*—Lester Brown's work, Joe Romm's *Lean and Clean Management,* and others: David Brower, and more recently the work of Dr. Karl-Henrik Robèrt of Sweden and the account of how he initiated The Natural Step. Since we've adopted The Natural Step, it's obvious we think it's important. I'll tell more about it later.

I devoured Herman Daly and John Cobb's book, *For the*

Common Good, trying to get a grip on the economics of sustainability. They propose a system of economics that recognizes external costs, such as the societal costs of greenhouse gases, and seeks means for internalizing them by, for example, calculating the true cost of oil and getting the price right. Their book proposes to turn traditional economics on its head by suggesting that the nominal economic decision unit, the enlightened, self-interest guided individual, be replaced by "persons *in community*" [emphasis added]; that decision making be based on premises such as a full world with physical constraints being pushed to their limits, rather than an empty world with no constraints, and a world of finite resources rather than infinite resources. Earth's vital signs suggest that these are timely ideas.

During that first year of learning and growing, I also met and came to know and love John Picard. John is consultant to the Southern California Gas Company's Energy Resource Center building project (the ERC). It's a landmark, "green construction" demonstration building, and the influences of Picard, Paul Hawken, and Bill McDonough can be seen all through it. Hawken's book brought John and me together, and I worked with John to devise something he and the Gas Company really wanted in this building, the first-ever in the history of the world (to my knowledge) perpetual lease for carpet.

We called it the Evergreen Lease. In the Evergreen Lease, Interface, as the manufacturer, not only made the carpet with state-of-the-art recycled content, but we also took responsibility for installing the carpet and maintaining it. Not only do we clean it regularly, but because it is free-lay carpet tiles, we selectively replace worn and damaged areas, one 18-inch square at a time. We implement a rolling, progressive, continuous facelift by periodically, over the years, replacing modules and, most importantly, recycling the carpet tiles that come up. We continue to own the carpet. Title for the carpet tiles never passes to the user, the ERC; it stays with us, the manufacturer, along with the ultimate liability for the used up, exhausted carpet tiles. The Gas Company pays by the month for color, texture, warmth, beauty, acoustics, comfort under foot, cleanliness, and healthier indoor air (Intersept is built in)—the services carpet delivers—and avoids the landfill disposal liability altogether; that's our problem, and we intend to convert that liability into an asset through closed loop recycling. We deliver the benefits and the services of carpet, but continue to own the means of delivery—theoretically for as long as the building stands.

Here's the thing: The economic viability of the Evergreen Lease for us *and its ultimate value to Earth* depend on our closing the loop. That is, we must be able to recycle *used* face fiber into *new* face fiber to be made into new carpet tiles, and *used* carpet tile backing into *new* carpet tile backing. We have yet to learn to do either economically or energy efficiently, much less to drive the process with renewable energy which is necessary for sustainability. So, you might say, we're cantilevered a bit and out on a limb of sorts. But we will get there. It's key to achieving sustainability, along with thousands of little things and the other big ones, those technologies of the future that put T (Technology) in the denominator, and other new ways of doing business yet to be envisioned.

The Evergreen Lease is a manifestation of the future, not just for carpet but for a wide range of manufactured durable goods. It's one example of how commerce can be redesigned for the 21st Century to use abundant labor to reduce dependence on diminishing virgin resources, and to increase efficiency in resource usage by forcing manufacturers to think responsibly cradle-to-cradle. We're grateful to the ERC and John Picard for driving this concept to a reality and letting Interface be a participant. I'll add this footnote: For the Evergreen Lease concept to become broadly successful, not only must we master closed loop recycling, but the financial institutions must get outside their comfort zones, too, and become third party participants, providing financial intermediation for this strange concept they never before encountered, a lease without a term for a product with indeterminate salvage value. Who can say what the value of those salvaged molecules will be in years to come? That will depend on the price of the fossil fuel precursors for virgin materials, which most likely will depend on revolutionary tax policies that are not yet on Congress' radar scope.

❖ ❖ ❖ ❖ ❖

During these four years, I've continued to read, even the other side's point of view. Early on, a friend took issue with me and disputed Hawken, Lester Brown, and others, calling them "alarmists." We had a friendly debate going for a while. He sent me Bast, Hill, and Rue's book, *Eco-Sanity*. It's the other view, that of the "foot draggers," one might say. It says good science doesn't support the views of the alarmists; that the world has 650 years' supply of petroleum, not 50; that the concern over the ozone layer is misplaced and unfounded; that acid rain is a disproven theory; that global warming is, too; that problems with automobiles, nuclear power, and oil spills are past problems that are nearly solved; that pesticides and toxic chemicals are manageable problems; and that deforestation and resource depletion are problems limited mainly to third world countries.

There's another book out there: *The True State of the Planet,* edited by Ronald Bailey, that conveys a similar, "the sky is *not* falling" message to "chicken little" environmentalists, so to speak. It forecasts a coming age of abundance, says we can wait a while on global warming to get the computer models perfected, claims that famine is a thing of the past for most of the world's people, and so forth. These people quote historical data effectively, put great faith in human intelligence going forward, and write persuasively. They will test your resolve. They shook mine at first.

Honest people of goodwill and with good intentions can disagree. Just as we have technophiles and technophobes, all with sound reasons for their positions, I suppose good people can even interpret the same data differently and reach opposite conclusions, without having to be branded as foot draggers or alarmists. But how do we reconcile all of this? Where's the truth? I struggled for a year to find some answers to these questions. Had I put my company on a hopeless, misguided tangent?

The title of my second outside public address on this subject, given to the U.S. Green Building Council at Big Sky, Montana in August 1995, was titled: "The Journey from There to Here: The

Eco-Odyssey of a CEO." Environmentally speaking, "There" was where I had been in August 1994, before that first task force meeting, pushing Intersept and IAQ through the Envirosense Consortium to make a buck, and staying in compliance on all the rest. "Here" was where I was one year later, speaking at Big Sky, with an awakened, sensitized conscience—realizing, for example, that compliance could mean "as bad as the law allows"—and with an awakened, sensitized company, hoping to do what's right, after wrestling for a year with what the truth was in all of this, and looking for a reconciling statement.

Well, I felt I had found that reconciling statement in my year of wrestling or, at least, the beginning of one. Here it is: Our planet is billions of years old and has billions of years to go; creation goes on. David Brower, the 85-year-old former Executive Director of the Sierra Club, often quotes son Ken Brower's observation that a living planet is a rare thing, perhaps the rarest thing in the universe. Also, thanks to very sophisticated modern technology that allows us to read the past, David has put *us,* our history as a species, our agricultural revolution, and our industrial revolution, into thoughtful perspective by compressing all of geologic time, from the initial formation of Earth 4.5 billion years ago right up to now, into the six days of biblical creation.

Using that compressed time scale (one day = 750,000,000 years), Earth is formed out of the solar nebula at midnight, the beginning of the first day, Monday. All day Monday is spent getting geologically organized. There is no life until Tuesday morning, about 8:00 a.m. Amazingly, life, beginning with that first spontaneous cell somewhere in the primordial oceans, lifts itself by its own bootstraps, and survives! The *prokaryote bacteria* appear quickly, then proliferate, into mind-bending diversity, ever more complex. About Tuesday noon the blue-green algae already begin to create the oil deposits.

Millions upon millions of species come during the week, and millions go. What begins as a very toxic and hostile environment gradually is detoxified and sweetened as each species, through its metabolic processes, prepares the hostile environment for the next species, and the next, gradually sweetening Earth's evolving biosphere and preparing the way for those that preceded us, and for us.

Thursday morning, just after midnight, photosynthesis— gradually building since Tuesday—gets going in high gear. Oxygen begins to accumulate in the atmosphere and the protective ozone shield begins to develop. Soon after, early Thursday morning in the wee hours, more complex *eukaryote* cells, like those that will come to make up our own bodies, appear. Life begins then to really flourish and evolve into more diverse and more complex forms.

By Saturday morning—the sixth day, the last day of creation— there's enough oxygen in the atmosphere and sufficient ozone shield in the stratosphere that the amphibians can come onto land, and there's been enough chlorophyll manufactured for the forests and other land vegetation to begin to form coal deposits.

Around four o'clock Saturday afternoon the giant reptiles appear. They hang around for quite a long time as a class of animals goes, until 9:55 p.m., nearly six hours. (That would be *really* long for a species. None has lasted that long, and our species is not likely to either!) Just a few minutes after they are gone, a bit after 10:00 p.m. Saturday night, the primates appear. (Incidentally, the Grand Canyon begins to take shape only about 16 minutes before midnight.) *Australopithecus,* the first species on that branch off the main primate branch, the one that eventually leads to us, shows up in Africa at 11:53 p.m., seven minutes ago. *Homo sapiens sapiens* arrives at 11:59:54—that is us! Arriving on the scene just six seconds ago! The last six seconds at the end of a very long week, that's how long we've been here.

"Let the party begin!" David Brower says. But the party becomes a binge when, just a little over one second ago, 1.2 seconds in geologic time, we (i.e., our forebearers) throw off the habits of hunting and gathering and settle down to become farmers, and begin to change and sacrifice the environment to suit, and feed, our appetites. A third of a second to midnight: Buddha. A quarter of a second to midnight: Jesus of Nazareth. One-fortieth of a second ago, the industrial revolution ushers in the age of technology; the party picks up steam, so to speak, and kicks off the great carbon blow-out that will characterize the first industrial revolution. An eightieth of a second ago, we discover oil and the carbon blow-out accelerates. One two-hundredths of a second ago we learn how to split the atom, and the party gets very dangerous, indeed. I would show a time line for this week, but the last one-fortieth of a second would not be discernible. If the time line were *one mile* long, the industrial revolution would occupy the last 0.003 inch! One human lifetime, about 0.001 inch.

And now it's midnight, the beginning of the seventh day. The Union of Concerned Scientists, numbering some 2,000 (including more than 100 Nobel laureates), told us in 1992 that we had *one to a few decades*" [emphasis added] to reverse course. In other words, the next two-hundredth of a second will be decisive; the time since we learned to split the atom (I remember 1939, so less than one lifetime), that short span of time projected not backward, but into future, will be decisive. God can afford to rest on the seventh day, but I do not believe we can. I believe Earth needs a miracle. To paraphrase Gandhi, who said, "If you want change, you must be the change," we must be that miracle.

To put it another way, the 10,000 years since the agricultural revolution began is, say, 500 generations. Fifty years of oil is two-and-a-half generations worth; 650 years is 32-and-a-half generations worth. Seems like a big difference on our scale of observation, but whether we're living in the last one-half percent of an epoch or the last six percent of an epoch doesn't really much matter. Time is short. In a blink of God's eyes, in an instant of geologic time, the whole epoch will be over. Whether the earth will run out of oil in 50 years or 650 years may seem like a big contradiction in conclusions reached, but either, in geologic time, is the snap of a finger. Just that quickly, the reindeer of St. Matthew Island take on deep and personal significance if we care about future generations.

Our life span is so short that it's like being in only two or three frames of a movie that has been running a long time and has a long time yet to run. Our time on Earth is just so brief that we don't see enough of the movie, can't even see the next scene, much less where it's all headed. But our few frames can have an enormous effect on the outcome of the movie. Not to trivialize through analogy, but I remember hearing a NASA scientist say once, talking about Apollo XI, that first man-on-the-moon expedition, that 90 percent of the time the spacecraft was off course. The critically important mid-course corrections made it possible to reach the moon, and that determined the outcome. I stand firmly convinced that Earth—no, humanity—is off course and desperately needs a mid-course correction.

Dear reader—if I may be so bold, fellow plunderer—we've done a lot of damage in one-fortieth of a second—with our technology—since the beginning of the first industrial revolution and, *to be sure*, created enormous wealth, prosperity and economic growth in our part of the world, the developed world, but at what price to Earth? At what price, measured in God's currency? What is God's currency? I don't know exactly. At Interface, we're trying to figure it out, but it surely is not dollars. Common sense tells us that neither the damage nor the economic growth—growing out of technology *as we practice it today*—can continue indefinitely; that the ways of the last one-fortieth of a second cannot go on and on. The first industrial revolution is unsustainable. We must not go on denying it. There is a limit to what our finite Earth can supply—and endure.

If it is true, as I was taught in college Economics 101, that all wealth ultimately comes from Earth, then it must follow that wealth creation at the cumulative expense of a finite Earth is not a sustainable process. What was taught in Economics, circa 1952, needs rethinking. Can it be true wealth if it is stolen from our descendants by the long arm of remote tyranny? If it is created by consuming Earth's capital reserves? Could we run our businesses or our households that way for very long, consuming our capital? This notion of wealth is more appropriately described as our children's inheritance, entrusted to us for safe-keeping, being squandered for today's excesses.

Human intelligence, through design, together with human labor and energy *from the sun through photosynthesis* (up to now coming mostly by way of Earth's stored fossil fuels), add form, order, and purpose to raw natural resources; that is, they add value—all value—that we then use up. The whole process of wealth creation, as it has been pursued for the most part, depends ultimately on raw natural resources, whether from the forest, the field, the mine, the oil well, or the ocean. The process of creating usable value in them is always driven by the energy of the sun, either stored or direct, but so far mostly from stored fossil fuels.

So, how much can we expect the earth to yield from its finite, solar-derived, stored resources? It is self-evident that there's a limit. Therefore, we must learn to create wealth from more efficient use of resources and, eventually, from utilizing current solar income, rather than living off the past until its stores are exhausted and denied to future generations. If we don't, we, i.e., our progeny, will surely go to the poor house—another way of saying ecological collapse. Human intelligence can take us a long way, but ultimately it will not be able to create something out of nothing.

Dwindling resources eventually will dwindle to zero, given enough time, and there's a lot of that yet to be for Earth, about five billion years or more, though surely much, much less for our species.

Exacerbating this whole issue is another lesson from Economics 101 which defines "The Basic Economic Problem" as the gap between what each of us has and what each of us wants, a gap that drives all economic "progress," because it can never be closed. That is, no matter what we have, we want more. That's human nature. More precisely, that's human nature in our culture, the culture that Daniel Quinn calls the "taker" culture. Compound that with an ever-growing population, with each person wanting and striving for more, and it becomes clear that we have to find new, sustainable ways to satisfy needs and wants, other than by taking and taking from Earth's limited capacity to provide from its stored natural capital, and other than by dumping our poison into her limited sinks. What could be more obvious from the perspective of geologic time?

We just have to begin where we are, not where we wish we were—sooner, not later, according to the scientists—to take those first steps in the long journey to sustainability, and begin to dismantle the destructive, voracious, consuming technologies of the first industrial revolution. We must start to replace them with the kinder, gentler technologies of the next industrial revolution, moving T from the numerator to the denominator and making technology part of the solution. We must begin to reinvent business, commerce, and probably this whole civilization (something Daniel Quinn is trying hard to think through), and find ways to create wealth (perhaps, redefine wealth), meet needs, satisfy wants, and raise standards of living for all without taking them out of Earth's hide.

Perhaps that has already begun as a tiny gesture when we beam somebody up on a video conference screen rather than get on an airplane to burn our part of the jet fuel that powers the plane, with us, from here to there. I say that's a start. It's one of the 10,000 little things we can do to conserve resources, avoid pollution, and honor nature. It's a beginning, but we really need to get busy identifying and doing the other 9,999. Conscience demands it. Or would we have our great-grandchildren curse us?

Another part of the reconciling statement lies in what I'll call the McDonough Paradox. Bill and I were talking one day about the contradictory positions of the two polarized schools that I've hesitantly called the "alarmists" and the "foot draggers." Let's express those positions in terms of *perception, action,* and *outcome.* The alarmist perceives Earth to be in crisis, sees our actions as totally inadequate, and predicts the outcome to be collapse. On the other hand, the foot dragger perceives things as not so bad, even getting better, sees our actions as good enough, maybe too good— meaning expensive and misguided—and sees the outcome as an abundant future for all.

Here's the paradox: the surest way to realize the alarmists' outcome, collapse, is to accept the foot draggers' view of where we are and what we need to do. On the other hand, the surest way to realize the foot draggers' outcome, abundance, is to believe the alarmists' view that we are in trouble and have to change. "Thesis and antithesis, reconciled through synthesis," to quote my friend, economist Mark Sagoff of the University of Maryland. He says Hegel would like that, furthering as it does the Hegelian view of the process of history.

Bill McDonough, himself, puts it this way: "You're the alarmist and you have a big bet with your foot dragger friend about how it will all turn out, and you're working like hell to lose that bet."

When I first became sensitized to the environmental crisis in 1994, it seemed to me that there was a pretty heated dialogue going on between those parties at opposite ends of the spectrum, the people I have called alarmists and foot draggers to dramatize their diametrically opposed views.

On one hand the alarmists seemed to be saying, "It's hopeless. Humankind is doomed, and when our species goes, millions more will go, too."

On the other hand, the foot draggers seemed to be saying, "Nonsense, things were never better. The rivers are cleaner, the air is cleaner. Living standards will continue to rise, world population will stabilize. Everything is going to be OK."

To try to reconcile these opposing viewpoints, I turned to the geologic time scale to understand what a relatively short time *homo sapiens sapiens* had been around and what shameful damage the species had done, especially in the last "one-fortieth second" since the industrial revolution began. The McDonough Paradox helped, too, to convince me time was short, the crisis urgent.

However, as I now listen carefully to the opposing arguments, I hear something different emerging. I hear the foot draggers saying that the *historical* record just doesn't support the claims of progressive deterioration becoming catastrophic degradation; that rivers no longer catch fire; that the particulate count in our cities' air is dropping. (That may be true in many places, but it's not true everywhere, as anyone who's been to Bangkok, Sao Paulo, Mexico City, or Beijing knows.) They say that proven oil reserves are increasing; that food supplies per person are increasing; that the alarmists just don't put enough faith in human intelligence, ingenuity, and survival instincts.

Meanwhile, the alarmists are saying that we've just got to do something; we can't sit on our hands; that tax policy is wrong at the governmental level; that we tax good things, such as income and property, things we would like to encourage rather than discourage by taxing them. Instead we should be taxing bad things, such as pollution, waste, and carbon dioxide production, things we should be discouraging rather than giving a free ride or even subsidizing. We could change that with some applied intelligence and political will. We could harness current solar income with applied intelligence and stop consuming stored natural capital. We could close the recycling loop and eliminate waste and toxic emissions with applied intelligence. We could change the way we live, redefine wealth, come together as communities—with applied intelligence.

Now the different views don't seem so diametrically opposed. They seem to represent degrees of difference. Both views put the burden on applied human intelligence to find solutions. It's just a matter of whose sense of urgency you adopt. It's not so much a question of which direction we go, but how fast.

If we superimpose on today's dialogue the reconciling geologic time scale, we must be appalled by the damage done by our species in such a short time. Whether we take 100 years to repair it, or 1,000 years, either is still a blink of God's eyes, but fix it we must. In that sliver of geologic time it could be over for us. We must apply human intelligence to the problem, whether it's creating those new technologies, or re-thinking taxation policy, or changing our life style. Leave everything to market forces? For sure, market forces can help fix things, if we first apply human intelligence to redress the market's indifference to externalities, and get the prices right so they cover the true, all-in costs, and get the incentives right, too, and introduce those rewards that will stimulate the innovations that add to the denominator in the equation:

$$I = \frac{P x A}{T_2}$$

In other words, let's be smart. Common sense tells us the sooner, the better; the sooner we begin, the more likely we are to avoid the abyss of St. Matthew Island.

CHAPTER THREE

DOING WELL BY DOING GOOD

I STARTED IN THE MIDDLE of my story at that conference in Boston. Now you've got the beginning of my EcoOdyssey. It is becoming an eco-epic. But what about the ending? Where will our few frames in this movie take us, out of crisis or deeper into crisis?

Crisis is an interesting word. The Chinese symbol for crisis is a combination of two characters: 危 DANGER and 机 OPPORTUNITY

In terms of these two components, *danger* and *opportunity*, I want to outline how we as a company are thinking about this crisis.

First the *danger:* I believe that Earth is damaged and hurting— badly. There is so much happening that it is just frightening.

Paul Hawken has laid out for us a litany of disasters that are happening all around us. To those, as if they weren't enough, we can add such cheery facts as:

• The decline of grain production in absolute terms worldwide;

per capita, in decline since 1984! Grain feeds us directly and indirectly through livestock.

- The decline of fish catch per capita—worldwide. In decline since 1987, our own New England fishing industry experiencing collapse is an up close, immediate example.

- Creeping death for the inland seas—choking on pollution; fish catch in the Caspian Sea is down by 99 percent! Is it a microcosm of Earth's oceans in the centuries ahead? The decades?

- One billion people looking for work, but can't find jobs. Another billion people already living in starvation conditions. Another billion on the fringe, hanging on by their fingernails. Half of Earth's people in serious trouble. We cannot escape the consequences of such inequity. So many interrelated factors— population, food, jobs, industrial production, standards of living, toxic waste, pollution, global warming habitat loss, species extinction—all interrelated in ways that, with present trends, point toward collapse for the human species. We cringe from the ravages of the AIDS epidemic and we wonder, what's next?

- The loss of the rainforests—at the rate of a football field every two seconds—NOW, NOW. Even while the Brazilian government says it's not so, the satellite photos say it is so. Millions of acres of Sumatra and Borneo's rainforests, lost to fires, attest to the facts. An estimated half of all of Earth's species live in the rain forests, on just seven percent of Earth's surface.

- Disappearing wetlands—wetlands that provide nutrients for the beginning of the food chain, and you know who's at the other end. Us! The wetlands also are the cradle of speciation. Even smelly swamps have a vital role.

- A nuclear waste cleanup that is off the scale in terms of both horror and cost, once estimated eventually to reach $300 billion. The unfolding Russian and Ukrainian situation could make this two times or three times larger. But nobody knows how to begin. Even if we learn to store the waste safely, what language shall we use for safety instructions for people 20,000 years from now? No language on earth is that old, none has survived that long—and plutonium is a 500,000 year problem.

- Greenhouse gases, global warming, stratospheric ozone depletion (increasing UV radiation, increasing incidence of skin cancer), rising global temperature, melting polar ice caps. The largest iceberg ever known broke loose from the Antarctic ice shelf in December 1995 (their summer). As it drifts into warm waters and melts, to the extent it was once over land, not floating and already displacing ocean water, ocean levels rise an infinitesimal bit. The cumulative effect over time—devastating! Now, the British Antarctic Survey tells us that the 8,000 square mile Larsen ice shelf is critically unstable and may collapse. Temperatures around the ice cap are rising five times as fast as the global average, and grass has been seen growing along the edges of the icy continent.

- The scientific debate about global warming and climate change is over when 2,600 atmospheric scientists from all over the world agree and only a handful hold out in skeptical disagreement. The debate is now political. The science is compelling. Global warming is real! The political debate can go on, but sea levels will rise in the 21st Century. How much? The most likely amount, about 20 inches, will be enough for a strong storm to put much of Florida under water. It is already too late, probably, to save parts of Florida and other low lying coastal plains. Some 9,000 square miles of the United States appear destined to be lost in just the next 100 years, absent massive dike construction. We would fight World War III before allowing a foreign aggressor to take 9,000 square miles of the United States! What *will* the 22nd Century bring, when atmospheric concentrations of carbon dioxide exceed two times pre-industrial revolution levels, where today's computer models stop? What *will* a "four times" world be like? So far, there is nothing in the offing, *not even the Kyoto agreement* when ratified as treaty, to prevent it from happening. Action is needed now, the earlier, the better. The longer we wait, the higher the concentrations will go into the zone of unknowable consequences.

- Toxic waste dumps, superfund sites, that defy known cleanup technologies.
- Acid rain, killing whole forests in Central Europe and, along with industrial, agricultural, and municipal pollution, poisoning lakes everywhere.
- Atmospheric ozone drifting from our cities to rural areas, adding to the forests' distress and adversely affecting crop yields—a major factor in determining whether China will be able to feed itself while industrializing. A China that cannot

feed itself will be everybody's problem.

- And, especially apropos businesses like ours, non-renewable resources being gobbled up at obscene rates and, for the most part, burned for energy and converted into carbon dioxide to exacerbate the greenhouse effect.

With every life support system comprising the biosphere stressed and in decline, some say we're killing Earth. No, I think Earth will survive. Life too will survive, I think, even if it is beaten back and reduced to the last *prokaryote* cell to start all over again, as it started with that first cell, on "Tuesday morning" 3.85 billion years ago. Yes, I think life will survive, and that we probably will too, for a time at least—but the loss of three-quarters or more of the species that share this planet with us, over the course of the next century, would constitute an unforgivable crime against our children and their children, limiting their life-sustaining options in ways that we cannot even begin to imagine.

Fraser firs once thrived in the higher elevations of the Great Smoky Mountains of east Tennessee. Today, 90 percent of them are gone, their remnants standing like ghostly pylons against the green backdrop of the mountain slopes; the survivors, as were the vanquished, ravaged by the *balsam woolly adelgid,* an exotic species of insect accidentally introduced from Europe. The United States Park Service, explaining the tragedy of the firs, concludes with this chilling observation: *An exotic species eventually exterminates its host; then it dies, too.* We should study what ultimately happens to exotic species, because it surely looks as if we are one, wreaking havoc on every living thing in our path.

Earth will be around for another five billion years or more, but meanwhile it is hurting. We are killing species by taking more than our share, fouling our own nest, and, in the process, diminishing the quality of life for our species and perhaps even dooming ourselves to extinction. That sounds harsh, but many thoughtful people believe this, the unthinkable, is true; St. Matthew Island *is* a metaphor for Earth. Perhaps, too, is the Fraser fir. It's only a matter of time.

That is the *danger.*

I'm often asked to define the business case for sustainability. How about, for starters: survival? Without sustainability, our descendants will watch society disintegrate and markets evaporate before their eyes. We cannot live without the life support systems of the biosphere any more than the other species can, and we continue to seriously over-stress those systems. The stress must stop for society, much less business, to thrive.

Now, what about the *opportunity?*

First, you must understand that I am focused on the tiny corner where I live and work. If I cannot make a difference there, I surely cannot make a difference anywhere else. I have thought long and hard, and strategically, about how to make a difference through Interface, and I have read a lot about what has gone wrong and what seems necessary to make it right. Once I understood what Rachel Carson started, I felt morally obligated to help advance her legacy. Once one understands this crisis, no thinking person can stand idly by and do nothing. Denial is alluring, even seductive, but once you get past denial, you know you must do whatever you can. Conscience demands it; psychological liabilities begin to accrue.

But who will lead? The question demands an answer. Who can do something about all of this? Paul Hawken says that it must be business and industry; that they (i.e., we) must take the lead. I look in the mirror, and I think he's right.

It's not the church, sadly. Though there are some encouraging signs that this could change, the church doesn't quite get it yet, even though in my church we sing:

This is my Father's world,
And to my listening ears
All nature sings and 'round me rings
The music of the spheres.

Too often, still, the church dogmatically helps perpetuate the myth Daniel Quinn identifies in *Ishmael,* the myth that Earth was made for man to conquer and rule. We have lived that myth, Ishmael reminds us, and sure enough there the earth lies at our feet, bloody, broken, and conquered. Good people with awakened hearts could change this.

It's not government. The government never seems to lead, it always seems to follow, waiting for the people to create the parade, though it has a vital role to play with taxation policy: increasing taxes on "bad" things and relieving taxes on "good" things. What if our income taxes were reduced and our gasoline taxes were increased, but in an overall revenue-neutral way? How much better off we and Earth would be! How much better still if the price of a barrel of oil, through taxation if necessary, reflected its true costs, including all of its externalized costs. Yet, at this writing, the subject of tax shifts is not on Congress' radar scope, much less the legislative agenda. I saw a great bumper sticker the other day: "If the people will lead, the leaders will follow." It's time for the people to step up and create that parade the politicians will run to get in front of.

It's not education, though education has a very important role to play in raising awareness and sensitivity in students, helping to draw the map to sustainability for all the disciplines, providing critical research to get the facts straight, and integrating the disciplines in a holistic vision of a new civilization. My school, Georgia Tech, is taking a leading role. I'll come back to what that is and how it came to be.

But if it's true that only business and industry can lead effectively and quickly, then how do you move the largest institution on Earth, when it's actually millions and millions of separate entities? I tell my associates at Interface that if we (not just I, but we) get together as a company and take the lead, we can set an example for the entire industrial world, by first examining and understanding some basic things that Paul Hawken has pointed out for us:

1. *What we take from the earth (those 1.2 billion pounds in 1995),*
2. *What we make and what we do to Earth in the making of it (products, stack gases, effluents),*
3. *What we waste along the way (waste in all its forms).*

What we take, what we make, what we waste—first to understand, then to do something about it, showing that it is good business to challenge and change all of these with Earth's benefit as the controlling criterion.

To expand the business case for sustainability, I believe that through EcoSense and QUEST we are pioneering a new business paradigm for success: "Doing well by doing good." There are two sides to that coin: 1) doing well (in a strict, money making, business sense), and 2) doing good. To be clear, we want to do good because it's the *right* thing to do. We have to get our hearts right. But doing good may, at first blush, seem altruistic, soft-hearted, soft-headed, even unbusinesslike. (Has Anderson gone 'round the bend? Well, yes, to see what's around there on the other side. That's part of my job. Having seen, I know I have never felt anything else that I have ever known in business to be, at once, so right and so smart. The closest thing was that first feeling when I fell in love with the idea of carpet tiles, a notable time when, once before, I went around the bend.) Who cares about that tree hugging stuff? Well, I do, and many (maybe most, or all) of our people either do or are beginning to. And our customers—especially the architectural and interior design communities—do. They generally want to do the right thing, and want to do business with companies that are doing the right thing. And a growing number of end user and original equipment manufacturer (OEM) customers do, too.

I believe we can *also do well* by doing good. After all, we went into business in the first place to do well. But how do we do *well* by doing *good*?

In three ways, I believe:

First, as I've already suggested, by earning our customers' goodwill and, hopefully, their predisposition to trade with us, to help us in this hard, hard climb. But to earn that goodwill we have to avoid "greenwash." Do you understand the term? Think of whitewash being used to cover a rotten fence. We must avoid that cynical, holier-than-thou, superficial cloak of green insincerity—so obviously self-serving, e.g., promoting products as "green" when they are not. We must be genuine. Our actions must speak louder than our words. Greenwash (pseudo-green) is, and should be, business suicide. Our customers should and will see right through it.

Second, through achieving resource efficiency. Amory Lovins, the brilliant physicist at the Rocky Mountain Institute who's working on that super-efficient hypercar, uses the automobile to illustrate the vast inefficiency of our industrial system. He says that the objective of the conventional automobile, which weighs about 4,000 pounds, is usually to deliver a cargo, averaging about 165 pounds, from Point A to Point B, maybe to pick up a one pound loaf of bread. What is the efficiency of that automobile? Lovins says that, with internal inefficiencies (feel the wasted heat from the engine block and the brakes), only 15-20 percent of the fuel energy reaches the wheels as traction. Most of that moves the car's weight; only a small fraction, the driver's. The net result: about one percent of the fuel energy moves the driver. The hypercar will increase that by a factor of 10—in time, more. So, what's the point of the traditional approach of spending millions to improve the internal combustion engine's efficiency by one or two percent? That's the wrong problem, and its solution will have a minor impact. The National Academy of Engineering agrees, and has concluded, using similar logic, that the overall thermodynamic efficiency of our American economy is (are you sitting down?) about 2.5 percent. Europe and Asia are not much better. The western economy is a waste machine, producing 97 percent waste!

When Interface takes petroleum or natural gas from the earth and, with the help of our suppliers, converts it to carpet which gets used once and goes to a landfill 10 to 15 years later, I think of that as equivalent to about 2.5 percent efficiency in the use of those precious organic molecules. When we have increased that efficiency a minimum of 10 times we'll still be only 25 percent efficient, with lots more room to improve. We are looking for those who will help us to do it. If you are one of our suppliers, or hope to be, take heed.

Where are we in this quest for resource efficiency? The $1 billion of sales we recorded in 1996 consumed 19 percent less material per dollar of sales than we consumed in 1995, reflecting both increasing efficiency and our shift toward services, especially downstream distribution. This happened while we were realizing record profits, which was not an unconnected coincidence. Cumulative progress over three years is an increase of about 22.5 percent in resource efficiency; our share price has tripled.

I'll say it again: I believe that in the 21st Century, the most resource-efficient companies will win! The sustainable will win big when oil's price finally reflects its cost and is $100, even $200 per barrel. Someday the market (and the economists) will wake up and the price *will* reflect the cost. That's the day for which we, as a company, are preparing.

At whose expense will the resource-efficient win? At the expense of the resource-*in*efficient; so I tell my people, we will win and Earth will win! The best win-win I can think of. Business, like technology, can emulate nature and eliminate the inefficient adapters.

The third way we can do well by doing good is by setting an example that other businesses cannot ignore. The target group to influence, *other businesses,* are also our customers or potential customers. If we do well enough through creating goodwill and becoming resource-efficient, to the point that we are kicking tail in the marketplace, then that is the example other companies will see and want to emulate. Maybe they will become converts *and,* hopefully, customers, too. Then a positive feedback loop, the snowball effect, will take hold. The more good we do, the more well we will do. We can do the most good by doing the most well. By doing well, the more good we will do through example. Then still others will see, and there will be that positive feedback loop—one of the few that is good for Earth.

I believe our customers will help us if they believe we are sincere. They will buy from us, even invest in our stock, and help us climb that mountain—even climb it with us. It is very important that we succeed in the climb, that the example be persuasively appealing to business and industry everywhere. Other companies will not look at us because they are benevolent or altruistic or philanthropic, nor because we are, but because we are undeniably succeeding in a different and attractive way. As our customers pitch in, our leverage with our suppliers will grow, as will our ability to bring them along on the climb. We cannot do it without both of them, customers and suppliers. Eventually, investors will see, too, that the only way for companies to do well in the long term is by doing the commensurate amount of good. Investors' capital speaks loudly, indeed!

That is the *opportunity*: To do well by doing good, and to make a difference by example—on a global scale—by making a difference in the corner where we live and work and inviting others to take a look and join in. Truly—to complete the business case for sustainability—entrepreneurs everywhere should thank Rachel Carson for starting it all; there are not only new, but noble fortunes to be made in creating and bringing to market the technologies, the products, and the services of the next industrial revolution. Is it good business? Reach your own conclusion (See Fig. 3-1, next page).

Of course, doing well by doing good and turning danger into opportunity are not the ending; they're only a tiny beginning of one possible ending. Who can say what the real ending will be? How long *homo sapiens sapiens* will make it? Will going beyond the limits of growth that Dana Meadows writes about in *Beyond the Limits*, in fact, soon lead to collapse and usher in an era of vast misery? Will self-declared wise, wise man be replaced in the "next few seconds" by a truly wise species?

It's safe to say, I believe, that if the developing world develops according to the model represented by the developed world, there will be unmitigated disaster in the 21st Century. It helps sometimes to think in analogous terms. Most people in our industry

FIG. 3-1

Interface, Inc. Net Sales and EBITDA*

net sales in millions of dollars
EBITDA in millions of dollars
EBITDA percent of sales

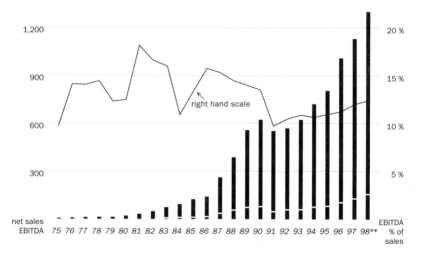

* EBITDA is earnings before interest, taxes, depreciation, and amortization
(the measure of profits that is most important to banks and other lenders).

** Analysts' consensus estimate

understand what is meant by "footprint" of a building, the plan view shape of the area a building occupies on its site. But we know that is not the entire footprint. There are also the mines from which the iron ore and other minerals came, the factories that produced the steel and glass, the transportation system that connects the building to its occupants, the power plant that supplies the building its electricity, the telephone company, and all the other supporting resources. When we think about it, they too are part of the building's footprint. Well, the footprint of a nation can be thought of as how much land it uses to meet all its needs. There are estimates that the United States already has a footprint larger than its geographical area. One indication of this is our balance of trade deficit. Let's say (though some think this very conservative) that the U.S. footprint today is 1.2 times its geographical area; but with three percent annual growth, we are on a course to double economically over the next 25 years.

At the same time, China, with a land area of about the same size as the contiguous 48 states of the United States, but with nearly five times the population, aspires to a standard of living for its people 25 years from now equivalent to ours today. What will its footprint then be? Well, something in the neighborhood of six times its land area. For example, if per capita automobile driving were the same in China as ours is today, China would consume the world's total gasoline production. So would India! And what about the rest of the developing world? Don't they, too, aspire for what the developed nations enjoy today, and how can we deny them their aspirations? And the rest of the developed world, like the United States, continues to increase its take, also. Two comments from world leaders come to mind.

Maurice Strong, founder of The Earth Council, former Chairman of Ontario Hydro, and Special Advisor to the Secretary General of the United Nations, has said, "The fate of the Earth will be decided in the developing world." I think he is only partly right. I think Earth's fate is also in our hands, because many of the solutions for the developing world must come from the developed world where the capital and expertise exist for developing those solutions. There are plenty of smart, creative people in the developing world, too, and they are watching our example as they decide how to deploy their more limited resources.

The second comment came from President Clinton. I was in the audience as he spoke at Georgetown University in October 1997, recounting how he had told China's President, Jiang Zemin, "The thing that I fear most is that China will get rich the same way we did in the West."

When the collective footprint of all the nations exceeds the area of all of Earth, something will have to give. We must get either another planet or a new model. The current western model that exploits Earth (leaving T in the numerator) won't do, and we don't have much time to fix it. How do we begin? Taking that critical next step, in what direction shall we move? Toward what end? What shall we use as a map, as a compass?

CHAPTER FOUR

A MOUNTAIN TO CLIMB

DEEP INTO HIS BOOK, *Ishmael,* author Daniel Quinn, speaking through the teacher, Ishmael, uses a metaphor to describe our civilization as it has arisen out of the first industrial revolution and the agricultural revolution before that. If you find yourself agreeing that maybe this civilization of ours is just not working out quite right, that something's seriously wrong, you will relate to Daniel Quinn's metaphor.

He likens our civilization to one of those early attempts to build the first airplane—the one with the flapping wings and the guy pedaling away to make the wings go. You've seen them in old film clips. In Quinn's metaphor, the man and the plane go off of a very, very high cliff and the guy is pedaling away and the wings are flapping, the wind is in his face, and this poor fool thinks he's flying. But in fact, he is in free fall, and just doesn't realize it because the ground is so far away. Why is his plane not flying? Because it isn't built according to the laws of aerodynamics, and it is subject, like everything else, to the law of gravity.

Quinn says that our civilization is in free fall, too, for the same reason: It wasn't built according to the "laws of aerodynamics" for civilizations that would fly. We think we can just pedal harder and everything will be OK; pedal still harder and even fly to the stars to find salvation for the human race "out there." But we will surely crash instead, unless we redesign our craft—our civilization— according to the laws of aerodynamics for civilizations that would fly.

In the metaphor, the very, very high cliff represents the seemingly unlimited resources we started with as a species and still had available to us when we threw off the habits of hunting and gathering, settled down to become farmers and, later, industrialists, and began to shape this civilization we have today. No wonder it took a while for the ground to come into sight.

But we are fortunate that there are people with better vision who have seen the ground rushing up toward us, perhaps sooner than most of us have; and others who have undertaken to discover those laws of aerodynamics for civilizations that would fly. Wendell Berry, David Brower, Paul Hawken, Daniel Quinn, Lester Brown, Dana Meadows, and others, carrying on the legacy of Rachel Carson, have seen the ground rushing up and have sounded the alarm for all to hear. Buckminster Fuller, Walter Stahel, Bill McDonough, Michael Braungart, Dr. Karl-Henrik Robèrt, and others have set out to define the laws of aerodynamics for civilizations that would fly: Fuller, recognizing that it's all a design problem; Stahel, McDonough, and Braungart with "waste equals food" and "cradle-to-cradle" cyclical design concepts; and Robèrt through his brilliantly conceived consensus document which laid out, for all to see, the science-based principles of sustainability which gave rise to The Natural Step. Let me tell you more about it.

Karl-Henrik Robèrt is a Ph.D. and an M.D. whose field is research oncology. He is highly respected in his native Sweden, especially for his work on cancer in children. Kalle (that's his nickname, pronounced kŏl'e) says that until about 1988, he had always thought that cancer was mostly a result of lifestyle, i.e., people brought it on themselves through self-indulgence. But when he saw an increasing incidence of cancer in children, he realized that children didn't have such lifestyles, so he began to believe that there were environmental causes. At the same time, he was struck by the contradiction between the way individuals, acting as parents, would do *anything* to help their stricken children, and the way those same people, comprising society and acting collectively, would do *nothing* proactively to prevent the same tragedies. It seemed to Kalle as if group intelligence sank below the level of the least intelligent individual in the group. He reasoned that this was because the group was without a shared framework.

So Kalle set about to move an entire society to adopt a shared framework, to see the connections, and to do something about it. He decided to try to create scientific consensus on the principles of sustainability to provide that shared framework. It's a wonderful story, well told in Robèrt's own publications, about how he came to achieve that consensus among a peer group of more than 50 of

Sweden's leading scientists, and then set about to educate a whole nation about sustainability. Among the scientists there was a great deal of disagreement, but in one area there was total, 100 percent consensus. This area of agreement was reduced to four fundamental principles. They have become recognized as the first order principles of sustainability. You might call them the consensus-derived, science-based laws of aerodynamics for civilizations that would fly— sustainable civilizations. Robèrt called these four principles the System Conditions for Sustainability. The "system" is the ecosystem, that thin shell of life where we and all the other creatures live, also called the ecosphere and the biosphere, that is 8,000 miles in diameter but only about 10 miles thick—from sea level, five miles down into the depth of the ocean, and five miles up into the troposphere. Relative to a basketball-size Earth, it is tissue-paper thin, and oh so fragile! For practical purposes, it sustains all life.

The principles of sustainability are based on scientific laws of nature that have been well understood for over a hundred years, the laws of thermodynamics. They are like the law of gravity. Someone has said, "They're not just a good idea, they are THE LAW—the law of the universe." Here are the first two laws of thermodynamics:

THE FIRST LAW *of thermodynamics says that* MATTER AND ENERGY CANNOT BE CREATED OR DESTROYED. *This is the principle of conservation of matter. When we burn something, it doesn't cease to exist. It changes form. When an automobile turns into a pile of rust, it doesn't cease to exist. It changes form. Every atom in the universe has always been in the universe. Every atom has existed since the beginning of time, and will exist until the end of time. It's true for matter; it's true for energy. Matter is energy. Neither can be created or destroyed.*

THE SECOND LAW *of thermodynamics says that* MATTER AND ENERGY TEND TO DISPERSE. *A drop of ink in a bathtub disperses. It may seem to disappear, but that's through dilution; it's still there, dispersed. Every manufactured article from the moment it takes its final form begins to disintegrate and disperse. A simple water glass, through the concentration of energy and design and human labor, is transformed from a pile of sand into a container, but from the moment of its completion it begins to disintegrate. If I drop it, I accelerate that disintegration. Another way to say it: the arrow of time flies in the direction of entropy, from order toward disorder. In a closed system, everything runs down. Everything that is concentrated eventually disperses.*

Matter and energy cannot be created or destroyed. Matter and energy tend to disperse. This means that any and all matter that is introduced into society will never cease to exist and will, sooner or later, find its way into our natural systems. It *will* find its way. It *will* disperse. Toxic materials are no exception. They, too, will disperse and find their way ultimately into our bodies. These are scientific principles. We can ignore them, but they will not go away. There are other laws of thermodynamics, but we can work with just these two for now.

Remember that process that started Tuesday morning (in geologic time)? Life lifting itself by its bootstraps, each species, through its metabolism, coupled with the sedimentation process, furthering the process of sweetening what started out as a toxic stew, preparing the way for the next species, and the next, and the next. It's important to understand that that toxic hostility has been, during the 3.85 billion years since "Tuesday morning," relegated to Earth's lithosphere—the crust. That's where it has been sent and sequestered through the inexorable process of sedimentation—down there; to make way for us, up here. Life has evolved in a self-reinforcing process: declining toxicity leading to more diversity, and more diversity leading to still less toxicity. New species after new species, a sweeter and sweeter Earth—and we are here today, the product of that self-reinforcing process.

What we have done in the one-fortieth of a second (in geologic time) of the industrial age is to reverse that process, to bring that stuff—lead, molybdenum, mercury, cadmium, antimony, copper, arsenic, asbestos, uranium, plutonium, hydro-carbons, etc.—back to the surface in a zillion different forms, right into our living rooms, so to speak.

Though there were at least three earlier mass extinctions, a look at the "recent" history of Earth (the last 100 yards of our mile-long time line), after amphibians came onto land and the evolutionary process was well along in the cleanup that prepared the way for us, reveals two cataclysmic reversals of the process. Something happened about 260 million years ago; scientists can only speculate about what environmental catastrophe created a "pollution spike." The spike was accompanied by a precipitous drop in biodiversity, a mass extinction, the most massive in Earth's history, as 96 percent of life vanished.

Then the forward evolutionary momentum was regained, and soon the giant reptiles evolved. Eventually they proliferated to rule the earth until another pollution spike occurred, almost certainly this time from a comet or an asteroid that struck Earth 65 million years ago near present day Yucatan and filled the atmosphere with toxic hostility. That time, more than 75 percent of life became extinct, including the dinosaurs, but for their distant relatives, crocodiles and birds. The good news is, mammals got a chance.

Today, in the last .003 inch of that mile-long time line that represents Earth's entire existence, another spike has surely begun, and, with it, another mass extinction. Seventy percent of those polled in the 5,000-member American Institute of Biological Sciences agree. This time, however, the cause cannot be attributed to unavoidable natural disaster, but to the deliberate, willful, and quite unbelievable action of the highest form of evolutionary creature yet produced, and to the fruit of that species' intelligence— the industrial age. As before, the new spike, still in its infancy, shows increasing toxicity and is confirmed by the precipitous plummeting of biodiversity, as species disappear faster than any time since the previous spike. Given the recent Asian oil discoveries, more efficient oil recovery technologies, and the extensive unmined deposits of coal, fossil fuels alone represent enough potential toxicity to kill us all if we foolishly continue to burn them for energy and dump their products of combustion into the atmosphere. With all the other "stuff" coming daily from the lithosphere and being created by humans, the toxicity spike could go very high, indeed, and the loss of biodiversity plummet a very long way.

FIG. 4-1

Self-Reinforcing Process

not to scale, for illustration only
in mya (million years ago)

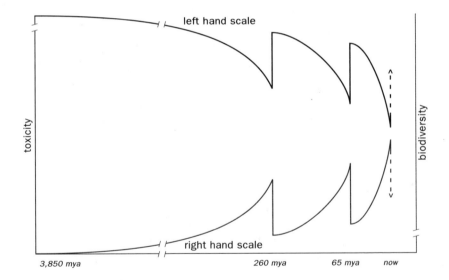

FIG. 4-1. Conceptually, the mutually reinforcing relationship between toxicity and biodiversity is represented by mirror image functions when the two variables are plotted against time. Though not to scale and disclaiming scientific precision, the graphs shown above depict the global trouble in which we find ourselves and suggest there's more trouble ahead.

Neither can we, as a species, adapt fast enough to tolerate such a drastic and sudden upheaval of the Earth's lithosphere. So children get cancer; older people get cancer, too, often from asbestos exposure or the like in their younger years. More subtle, but nonetheless devastating, things happen, too. Sperm counts decline, putting whole species, including ours, in long-term jeopardy. Likewise, pesticides (man-made, unnatural substances) kill people as well as insects; PCBs collect in fish and in human mother's milk, rendering both unsafe to ingest and putting endocrine systems at risk.

The self-reinforcing process can flip-flop and go the other way, too, and that has already begun, thanks to us, the human race. Increasing toxicity leads to loss of species. Loss of species (trees, for example) leads to increasing toxicity, and still more species are lost. The "canaries in the mine" are warning us very clearly. We, as an endangered species, should be listening. Perhaps our endangerment stems from our unique ability among the species to engage in denial. Or is it simple arrogance that leads us to believe that it cannot happen to us? If we re-create the atmosphere of 500 million years ago, we will not survive.

With that for background, here are the scientifically derived (based on the laws of thermodynamics) principles of sustainability, the system conditions of The Natural Step, published by Kalle and his consensus reaching peer group, and ratified by American peer review as well. Please read carefully; this is what sustainability requires—absolutely, incontrovertibly. You cannot negotiate with a cell and tell it not to worry, that everything will be okay. Denial cuts no ice. It helps, too, to keep in mind the verity of geologic time. Here goes (and I quote extensively from the paper, "A Compass for Sustainable Development," by Karl-Henrik Robèrt, Herman Daly, Paul Hawken, and John Holmberg):

1. *Substances from Earth's crust must not systematically increase in the ecosphere.* This means that fossil fuels, metals, and other minerals must not be extracted at a faster pace than they can be redeposited and reintegrated into Earth's crust, turned back into nature's building blocks. If substances from Earth's crust systematically and inexorably accumulate, the concentration of those substances in the ecosphere will increase and eventually reach limits. We don't know what the limits are, but beyond the limits, irreversible changes will occur—so much radioactivity that we all die, so much lead in the water that we all become sterile, so much carbon dioxide in the atmosphere that the polar ice caps melt.

2. *Substances produced by society (man-made materials) must not systematically increase in the ecosphere.* This means that man-made materials must not be produced at a faster pace than they can be broken down and integrated back into the cycles of nature, or deposited into the earth's crust and turned back into nature's building blocks. If persistent man-made substances systematically and inexorably accumulate, the concentration of these substances in the ecosphere will increase and eventually reach limits. Again, we don't know the limits, but beyond those limits irreversible changes will occur. At some point, dioxins

kill. Enough dioxins will kill us all. At some point, DDT, DDE, DES, mercury-containing compounds, and PCBs begin to disrupt endocrine systems. Endocrine systems keep our species going, and keep other species going as well.

3. *The productivity and diversity of nature must not be systematically diminished.* This means we cannot over-harvest or reduce our ecosystems in such a way that their productive capacity and diversity systematically diminish. We must certainly protect the small fraction of species that are capable of photosynthesis. We must not cut down the forests. They produce the oxygen that keeps us alive. Our health and prosperity depend on the capacity of nature to reconcentrate, restructure, and reorder building blocks into new resources. Rainforests and fisheries, farmlands and aquifers must not be pushed beyond their ability to recover. Species must be preserved; diversity in nature, protected. Why? Because we simply don't know all the interconnections in the web of life, but we know we are part of that web. It is foolish to say that we don't need this, or we don't need that. This and that lead to us.

4. *Therefore,* [emphasis added] in recognition of the first three conditions, *there must be fair and efficient use of resources to meet human needs.* This means that basic human needs must be met in the most resource-efficient ways possible, and meeting basic needs for all must take precedence over providing luxuries for a few. Otherwise, we will reap a harvest of social as well as environmental instability. If people living in wooded or forested areas cut down all the trees for firewood because they

don't have another source of fuel, all humanity suffers from the loss of biodiversity, and from the erosion, climate change, flooding, and desertification that follow. Fair is one thing; efficient is another, but they are intimately connected. How can we lift the lowest economically without dragging down the highest? The answer lies in resource efficiency.

Resource efficiency is the rising tide that will float all the boats higher. If the National Academy of Engineering estimate is anywhere near correct in estimating that the overall thermodynamic efficiency, i.e., the efficiency in the use of resources in the U.S. economy to meet human needs, is 2.5 percent, then I say it again: Surely, we must create for ourselves and show the developing world a better model than that. Technology (T) must move to the denominator, and we must remove extracting from the dictionary definition of technology! Technology must help put a billion unemployed people to work in gainful employment, conserving precious resources through cyclical, renewable processes.

The Natural Step is about reorganizing businesses and communities to conform to these four systems conditions for sustainability. These system conditions will define sustainability when we get there. But of course, we are not there; we have a long way to go. How in the world do we get there when we, as a civilization, are headed in the other direction?

As we reach for these system conditions, our organizations must *systematically decrease* their economic dependence on underground metals and fuels and other minerals. Our organizations must *systematically decrease* their economic dependence on the production of persistent, unnatural, man-made substances. Our organizations must *systematically decrease* their economic dependence on activities which encroach on the productive parts of nature. And our organizations must *systematically decrease* their economic dependence on the use of unnecessary amounts of resources in relation to added human value, i.e., they must systematically move toward fair and efficient use of resources to meet all human needs, and put people to work to raise their standards of living, too.

This is The Natural Step. Karl-Henrik Robèrt created it in Sweden; Paul Hawken has brought it to the U.S., Jonathon Porritt, to the U.K. Interface has publicly committed to be a Natural Step company. That means we have adopted these principles as our compass in our search for the path to sustainability, as our shared framework for understanding what sustainability is. Others (companies and individuals) are joining up, too. The Natural Step is becoming a force for good.

This is hard stuff. These are unrelenting principles, and they will not go away. Today, we are violating every one of them in ways that must not go on. What about you? The laws of thermodynamics are undeniable, but no law says we must follow the principles of The Natural Step. That's a matter of choice. However, The Natural Step is telling us, in undeniable fashion, that we must, for the sake of humankind's future, replace T_1 with T_2 and move T to the denominator so:

$$I = \frac{PxA}{T_2}$$

It is telling us further, in equally undeniable fashion, that social equity must be part of the fabric of sustainability.

Have I made the case for the next industrial revolution? For the urgent need to supplant industrialism, as we have grown up with it, with a better system? Why would I, along with so many others, say that the first industrial revolution was a mistake and is unsustainable? If that still seems completely outrageous, let me try a different tack.

Dana Meadows is one of the smartest people I know—an expert systems analyst, author, syndicated columnist, college professor, and farmer. Dana has published an elegant paper, "Places to Intervene in a System." In the style of a David Letterman Top 10 list, she has listed these places to intervene in increasing order of effectiveness, beginning with number nine and working down to number one, the most effective place to intervene. Number nine on the list is to adjust the numbers, more of this, less of that. Working down the list, you find such things as adjusting the regulating negative feedback loops, driving positive feedback loops, and changing the goals of the system. Number one on the list is challenging the mindset behind the system—the paradigm, the perception of reality, the mental model of how things are—that underlies the system in the first place. Dana says that this is the most effective place to intervene, but also the hardest.

Now, we have systems all around us: Our transportation systems, our communication systems, our computer systems, our production planning systems, our systems of government, our accounting systems, our educational systems, our systems for managing our households, our regulatory system, our banking system, and . . . the industrial system that has arisen out of the industrial revolution.

What is the paradigm behind the modern industrial system? If you look at how it operates, you know it originated in another day and age, and it still views (or acts as if it views) reality as it did then:

- *Earth is an inexhaustible source of materials (natural resources). We'll never run out. There will always be substitutes available.*

- *Earth is a limitless sink, able to assimilate our waste, no matter how poisonous, no matter how much.*

- *Relevant time frames are, maximum, the life of a human being; more likely, the* working *life of a human being; sometimes, especially in business, just the next quarter.*

- *Earth was made for humans to conquer and rule;* homo sapiens sapiens *(self-named "wise, wise man") doesn't really need the other species, except for food, fiber, and fuel, and maybe shade.*

- *Technology is omnipotent, especially when coupled with human intelligence, specifically, left-brained intelligence (practical, objective, realistic, numbers-driven, results-oriented, unemotional); these will suffice, thank you very much.*

• *And (how about this one?) Adam Smith's "invisible hand" of the market is an honest broker.*

Paul Hawken's *The Ecology of Commerce* and other books I've read since, together with my own late-blooming common sense, have convinced me that every element of that paradigm is wrong, dead wrong, and that survival of our species depends on a new industrial system developing quickly, based on a new paradigm—a new and more accurate view of reality, one that acknowledges:

• *Earth is finite (see it from space; that's all there is!), both as a source (what it can provide) and as a sink (what it can assimilate and endure).*

• *There will come an end to the substitutes that are possible. You cannot substitute water for food, air for water, food for warmth, energy for air, air for food. Some things are complementary.*

• *Relevant time frames are geologic in scale. We must, at least, think beyond ourselves and our brief, puny time on Earth—so brief—and think of our species, not just ourselves, over geologic time.*

* *Man was made for Earth, not the other way around, and the diversity of nature is crucially important in keeping the whole web of life, including us, going sustainably over geologic time.*

* *Technology must fundamentally change if it is to become part of the solution instead of continuing to be the major part of the problem. T_1 must be replaced by T_2, and T must move from the numerator to the denominator.*

* *The right side of the brain, the caring, nurturing, artistic, subjective, sensitive, emotional side (in business, the "soft side" of business), is at least as important as the left side, perhaps a good bit more important since it represents the human spirit.*

* *The market is opportunistic, if not outright dishonest, in its willingness to externalize any cost that an unwary, uncaring public will allow it to externalize. It must constantly be redressed to keep it honest. Does the price of a pack of cigarettes reflect its cost? A barrel of oil?*

Though the proximate (indeed, immediate), life threatening problem we face as a species is the deterioration of the life support systems that comprise the biosphere, there is a problem behind that problem: the industrial system that has arisen out of the mindset, the flawed view of reality, that underlies the system. So, it must follow that this flawed view of reality is the ultimate problem. Survival of our species, therefore, depends most of all on changed minds. (Sound familiar? That's Daniel Quinn's point, too.) An industrial system based on an erroneous view of reality has no foundation and will crash and take us with it, like a civilization that is ignorant of the "laws of aerodynamics."

So at Interface, we have chosen to intervene in the system. We have chosen Dana Meadows' most difficult and most effective place to try to make a difference. The reinvention of Interface reflects the new and more accurate view of reality, a new mindset for a new industrial system. We are going about this reinvention ambitiously, aspiring to become the sustainable corporate model for the next industrial revolution.

Pioneering the next industrial revolution is a tall order. So is reinventing civilization. But someone must. What Rachel Carson started must continue. Either we, as a species, do it and find harmony with nature, or nature will whittle us down to size. She is in charge, just as she was at St. Matthew Island. Earth's resources *are* finite; Earth's capacity to endure abuse *is* limited. We must find a way to acknowledge and respect those limits.

❖ ❖ ❖ ❖ ❖

The journey, with so far to go, goes on. In a global economy of $40 trillion ($40,000 billion), it's very presumptuous, isn't it, for a little company headquartered in Atlanta, that started from scratch in 1973 and took 24 years to reach one billion in sales, to think it can intervene in so large and complex a system as industrialism itself? What makes us think we can make a difference on a global scale and on an issue of such overwhelming magnitude? I really don't *know* that we can, but Paul Hawken, John Picard, Bill McDonough, Michael Braungart, David Brower, Daniel Quinn, Amory Lovins, Hunter Lovins, Bernadette Cozart, Dana Meadows, Herman Daly, Lester Brown, Rachel Carson, Kalle Robèrt, and others have convinced me that we should try. Unless somebody leads, nobody will.

At the very *least* we will give our people and our company a higher cause and a long range reason for being. Abraham Maslow, describing the hierarchy of human needs, says a higher cause is important, and I agree. After compensation to meet their needs, according to Maslow, people want the opportunity to develop and grow personally and professionally. When compensation is sufficient and growth opportunity is satisfied, people want to work for a company that makes a difference, that serves a higher cause. At Interface, we have learned in a very practical way that the quest for sustainability, for the welfare of our children's children, is a powerful, binding force. It is that higher purpose.

At the very *most* (let's dream a little), we'll start a trickle that will influence others—maybe you—to start their trickles; when those trickles come together into rivulets, and rivulets become streams, and streams, rivers, something good can happen.

As we used to sing in Sunday School when I was a child, "Brighten the corner where you are, brighten the corner where you are." Many years after learning that Sunday School song, I was exposed to the writings of the 18th Century philosopher, Immanuel Kant, and his somewhat more sophisticated corollary, "The

Categorical Imperative." What a great cause in which to invoke Kant and his 18th Century admonition! To paraphrase: "Before you do something, consider what the consequences would be if everybody did it." If we all succeed, individually, in doing good for Mother Earth in the corners where we live and work, in setting the example for others, and in governing our actions by the Categorical Imperative, "What if everybody did it?" we will be helping Daniel Quinn in his mission to change six billion minds, creating a river of change, and giving Earth that mid-course correction. I believe it's not an option. It is humanity's only hope.

I have used another simile to describe sustainability as a mountain to be climbed. Let me expound. I have this mental picture of a mountain that is higher than Everest. It rises steeply out of a jungle that surrounds it. Most of us, people and companies, are lost and wandering around in that jungle, and don't know the mountain exists at all. Rather, we are preoccupied with the threatening, competitive "animals" all around us. A few have sensed the upward slope of the mountain's foothills under their feet. Still fewer have decided to follow the upward slope to see where it leads. And a very few are far enough along to have had a glimpse of the mountain through the leaves of the trees, to realize what looms ahead and above. Very few indeed have set their eyes and wills on the summit.

I am thankful that the people of Interface are in that small group. We have found the mountain's seven faces (Chapter One) that must be scaled. Moreover, we have a compass, The Natural Step, to help us stay focused and on track to the summit, and we're developing EcoMetrics to measure our progress. We have some wonderful mapmakers, our team of advisors (Hawken, McDonough, Picard, Quinn, Brower, Porritt, Robèrt, Stahel, Browning, Lovins, Lovins, Meadows, and Cozart), who are helping us find the finger-holds and the footholds to stay on the path to the summit.

What will the view from there, from the vantage point of sustainability, be like? I believe it will be wonderful beyond description, and I hope to see it before I die. I also hope others, still lost in the jungle or just becoming conscious of the upward slope, even beginning to explore it, will hear our cries of joy through the foliage and rush ahead to follow our path, someday soon to join us at the summit (or better yet, beat us there). There's room there for everyone, and certainly anyone headed in that direction welcomes the companionship, as we seek to create the prototypical company of the 21st Century, of the next industrial revolution.

"What's that?" you ask.

CHAPTER FIVE

THE PROTOTYPICAL COMPANY OF THE 21ST CENTURY

WHAT WILL IT LOOK LIKE, the prototypical company of the 21st Century that I want Interface to become—this model for the sustainable enterprise of the next industrial revolution?

Some time ago, I was watching the movie *Mind Walk*, based on Fritjof Capra's book *Turning Point*, about the interconnectedness of all things. Of course, Capra was talking about interconnectedness at the subatomic particle level, and that was my first exposure to the idea. I've continued to read on that subject, but the movie started me thinking on another level about Interface's interconnections, that is, the linkages between Interface and its constituencies—how some were good and others were bad, and how the good ought to be strengthened and the bad, eliminated, and how still other linkages should be added. This led to a series of schematics to help all our people understand how we are approaching this monstrously difficult climb.

FIG. 5-1. Here's the Interface corporate logo. The circle represents the whole world. The "I" represents Interface within that global context, as well as, in a more subtle way, its focus on commercial *interiors*. I laughingly tell my audiences that it looks remarkably like the on/off button of a computer, too, which raises the unintended question, "Are we computing here?" Internally, we think and talk a lot about what belongs inside that "Circle I," i.e., what constitutes Interface?

FIG. 5-2. A typical version of what goes inside (this applies to any company) is people, capital, and processes. Economists often put "technology" where I have put "processes." To my mind, "processes" is the broader word and the better choice. At the core are the company's values. The four elements vary specifically from company to company, but the general pattern holds for all. All companies, with their manifold distinctive differences, fit this general pattern.

But of course, no company stands alone like this. Any company is connected to some important constituencies.

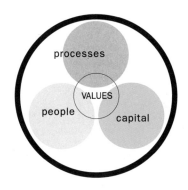

FIG. 5-3. In our case, Interface is part of a supply chain, with suppliers and customers and a market, our share of which we hope to increase. Products flow through that supply chain in one direction; money flows in the other direction.

But the supply chain doesn't stand alone either. It is connected to some other important constituencies.

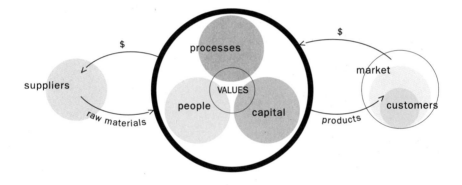

FIG. 5-4. Our suppliers are dependent on Earth's lithosphere for organic and inorganic materials. A very small amount of our raw material is natural, coming from the biosphere. Our processes are, unfortunately, connected to Earth's biosphere by the waste streams and emissions we produce. And the products we make end up too often, at the ends of their useful lives, in landfills, or worse, in incinerators, creating a further pollution load for Earth's biosphere to digest. Carpet in a landfill will last 20,000 years.

We are connected to our community, too. Our people come from there, and their wages return to the community's economy; often they are its lifeblood. Our capital comes from a sector of the community, the financial sector; if we are fortunate enough to earn sufficient profits, dividends and capital appreciation are returned to those investors, along with interest to our lenders. Government is part of community, too. We are connected to it through laws, regulations, and, of course, the taxes we pay.

With these linkages in place, we have a description of many, many companies, in fact, almost every manufacturing company on Earth and, by analogy, many other businesses and organizations. I have called this the Typical Company of the 20th Century. If this is all there is to Interface, Interface, too, is just typical.

However, we are trying to transform Interface into something different, a sustainable industrial enterprise. I call that enterprise the Prototypical Company of the 21st Century. While the transformation goes on in every business throughout our company, Interface Research Corporation, the people who started the EcoSense effort within our company by convening that first task force meeting in August 1994, is leading us in the process of transforming our company. Let's see what that means, step by step. How do we get there from here, and, in the process, pioneer the next industrial revolution?

Typical Company of the 20th Century

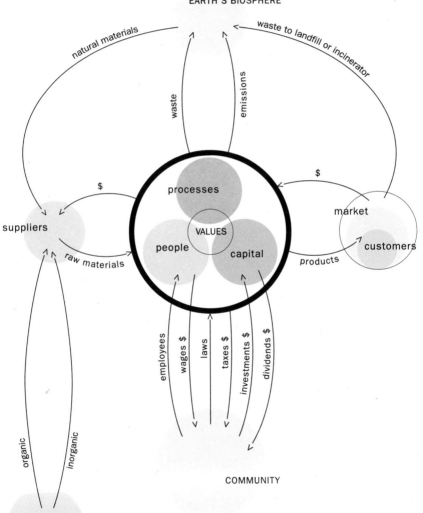

FIG. 5-5. We are pursuing the goal of creating the Prototypical Company of the 21st Century simultaneously on seven fronts (the seven faces of our "mountain"). Though we are at different stages with each, we hope all will meet at the top. The first one is shown in this figure. It is the front of Zero Waste. In pursuit of zero waste, to attack unwanted linkages to the biosphere, we have launched the effort we call QUEST. In QUEST, any waste is bad, and anything we don't do right the first time is waste. Against ideal operational standards—zero waste—we identified $70 million in waste, based on 1994 operations—10 percent of sales! We set out in 1995 on a mission to cut that in half by the end of 1997, then in half again by 2000 with hundreds of active projects, summed up and represented by each "X" in the diagram.

The PLETSUS listing in the Appendix describes our approach. We targeted $66 million of savings cumulatively over the first three-year period and much more in the ensuing years. We did not get all the way there. Actual progress is shown in Fig. 5-5a. But when we get to zero waste, the savings will be much greater because our company will have continued to grow, and the opportunity will have grown. Further, as we redesign our products to use less and less material and to last longer and longer, we are de-materializing the business and reducing the load on the biosphere at the end of the supply chain.

1 Zero Waste (through QUEST)

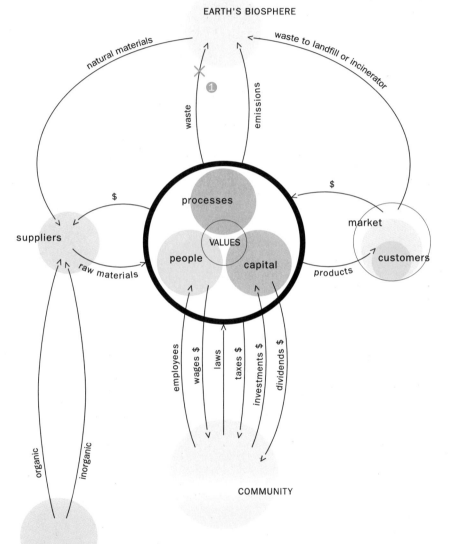

FIG. 5-5a. We didn't reach our goal of an index of .50 in 1997, but progress through fourteen quarters resulted in an index of .60, representing a 40 percent reduction in waste, saving cumulatively $67 million. This is real money, hard dollars, and it is paying for the rest of this revolution in our company. One quick result: scrap to the landfills is down over 60 percent from 1994 levels throughout our company, 80% in some operations.

We have reframed QUEST for the next three years, through 2000, and we find, even after the savings that have already been achieved, that with a larger company there is now $80 million in waste. Once again, we intend to cut waste in half in three years, then in half again and again, until all waste is driven out of our company and the concept of waste is eliminated. That will require reinvention over and over again. We must be a learning company.

FIG. 5-5a

QUEST Cumulative Results

index
cumulative savings in millions

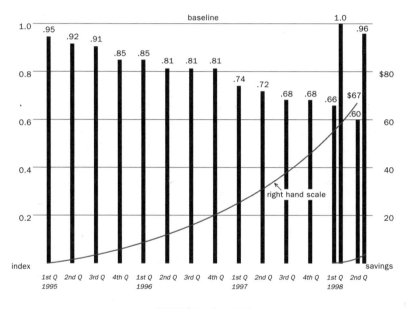

QUEST I baseline 1994
QUEST II baseline 1997

FIG. 5-6. The second front is Benign Emissions, to attack another unwanted linkage to the biosphere. We have inventoried every stack and every outlet pipe in our company, to see what is going out and how much of it there is, and we are reducing emissions daily. We have identified the world's most stringent regulatory standards and adopted them everywhere we operate. We began in 1995 with 192 stacks; with acquisitions since, that has become 229. Today we are at 184. There were 18 process effluent pipes; now there are 15. In all, 48 stacks and pipes have been closed off. I hope to live to see the last stack and pipe closed off, in factories that don't need outlet pipes for their cyclical processes. Again, many projects are represented by each "X."

But we know that to prevent *toxic* emissions altogether we must go upstream and prevent toxic substances from entering our factories in the first place. What comes in will go out, one way or another. We are just beginning to understand how difficult that undertaking is. Commercially available raw materials are replete with substances that violate the first and second principles of The Natural Step (see Chapter Four). Screening them out and remaining in business is a monumentally complicated undertaking. Yet we must. End of pipe solutions are unsustainable; they don't satisfy the principles of The Natural Step. Filters only concentrate the pollution, and then what do we do? We can't throw the filters away. There is no "away." Nothing is destroyed (first law of thermodynamics). It will disperse (second law of thermodynamics). Forget "away"; there is no such place to throw anything. Stopping pollution upstream is what we must do, leaving the toxic stuff in the lithosphere where the process of evolution, over the 3.85 billion years since "Tuesday morning," put it to make way for us. It must be left there (first system condition). Bill McDonough says that we must move the filters from the ends of the pipes to our brains, and focus our brains upstream to redesign products and processes.

2 Benign Emissions

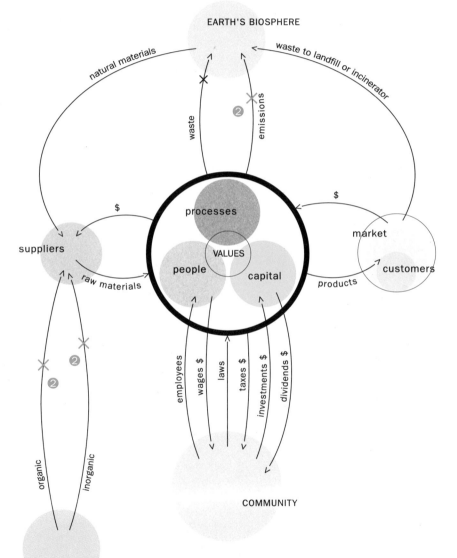

EARTH'S BIOSPHERE

natural materials

waste to landfill or incinerator

waste

emissions

②

$

processes

$

market

suppliers

VALUES

people   capital

customers

raw materials

products

employees

wages $

laws

taxes $

investments $

dividends $

organic

inorganic

②

②

②

COMMUNITY

EARTH'S LITHOSPHERE

FIG. 5-7. The third front, Renewable Energy, means eventually harnessing solar energy. In the short term, maybe hydrogen fuel cells or gas turbines, maybe wind (a form of solar), will run our processes, but eventually we believe it must be photovoltaic (pv.) generated electricity. Harnessing renewable energy will attack numerous unwanted linkages, both to the lithosphere and to the biosphere, and will allow closed loop recycling, the next front, to produce a net resource gain by obviating the need for fossil fuels for the energy to drive the recycling process. We have declared all fossil fuel-derived energy to be waste and targeted it for elimination under QUEST—two fronts, hooked up! The initial emphasis is on efficiency. Amory Lovins' principles are our guidelines, and he is our mentor. Only when energy usage is at its irreducible minimum are we likely to be able to afford the investments in renewable sources. How far can we go? Further than we ever imagined! In one case, through resizing pumps and pipes, we made a 12-fold reduction in connected horsepower for a key production line.

Our first application of photovoltaic power was in our Intek factory in Aberdeen, North Carolina. It is a nine kWp (kilowatt peak) unit that runs one 10 hp motor at a cost of 32¢ per kWh. (The cost is primarily depreciation on the capital investment.) A better use of the pv. power is to peak shave electrical demand during the hottest part of the day when the air conditioning load is greatest, and realize an effective cost of 15¢ per kWh, still four times the cost of fossil fuel electricity. Because it's not cost effective, the pv. array is a symbolic token. Greater savings are coming from natural daylight reflectors that track the sun from horizon to horizon to light the plant with daylight; the tracking is driven by a fraction of the pv.-generated power.

Yet, we are pressing on. The next pv. project is a 127 kWp unit in southern California to produce the world's first Solar-made™ carpet. We think Solar-made will sell, that our specifier customers will love the idea. Who cares if the energy costs a little more, if the product sells and helps Earth even a tiny bit? We're doing well and doing (a little bit of) good.

In Canada we have contracted with Ontario Hydro for "green power" (solar and wind). Even though it costs more, it's the right thing to do. We think Solar-made carpet will sell in Canada, too.

3 Renewable Energy

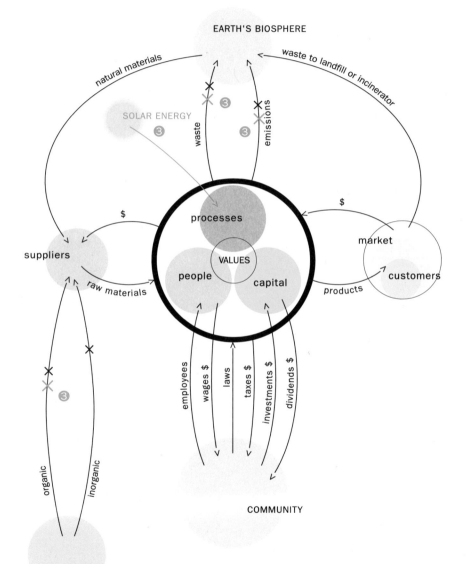

EARTH'S BIOSPHERE

natural materials

waste to landfill or incinerator

SOLAR ENERGY ③

waste

③

emissions ③

processes

$

market

suppliers

$

VALUES

people

capital

customers

raw materials

products

③

employees

wages $

laws

taxes $

investments $

dividends $

③

organic

inorganic

COMMUNITY

EARTH'S LITHOSPHERE

FIG. 5-8. The next front is Closing the Loop, to introduce closed loop recycling. Look at the impact this has on unwanted linkages! And see the new linkages that come into being. Two cycles are introduced: a natural, organic cycle, emphasizing natural raw materials and compostable products ("dust to dust") and a technical cycle, giving man-made materials and precious organic molecules life after life after life, through closed loop recycling. The "sustainability link," the part of the technical cycle where closed loop recycling will happen, must be invented and developed. It will be difficult and expensive to do, and we cannot do it alone. We need our suppliers' help here most of all.

But look at the power of it! The supply of recycled rather than virgin molecules in the technical loop, analogous to the supply of money in an economic system, will affect directly the resource-efficient "prosperity" of the enterprise. *What if everybody did it?* It would provide that rising tide that would lift the lowest on the economic scale, because recycling is labor intensive. Labor for natural resources is a good trade-off that will get better as the prices (of petro-resources) get right.

This front goes hand in glove with the previous front, renewable energy. What's the gain if it takes more petro-stuff to create the process energy than is saved in virgin raw materials by recycling? Two more fronts, hooked up! If we can get both right, we'll never have to take another drop of oil from the earth. That's the goal. It epitomizes our vision, along with factories with no outlet pipes; except, unavoidably, the next front stands in the way.

4 Closing the Loop

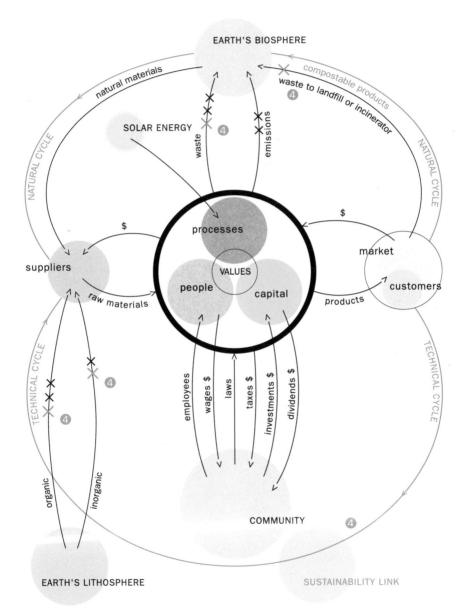

FIG. 5-9. Resource-Efficient Transportation, the fifth front, is the front that is least within our control and the hardest for us to crack, especially with 100 percent sustainability as the ultimate goal. We can video conference to avoid the unnecessary trip for a meeting, and we can drive the most efficient automobiles available. We can site our factories near the markets they serve, and plan logistics for maximum efficiency. But unless we choose to shut down contact with our customers and go out of business, we are dependent on the transportation industry for this one. Isn't everybody?

The good news is, progress is being made—with electric cars, hybrid gas/electric cars, jet engines powered by hydrogen (coming from biomass or, someday, water), and hydrogen fuel cells that are advancing in efficiency and cost reduction. Peter Russell's "global brain" (see his book, *The Global Brain Awakens*) is waking up, and the transportation industry is part of it. We need more "alarm clocks" to speed the process. Be one! We need Amory Lovins' hypercar. At the end of the day, we will have to resort to carbon offsets to completely resolve this one. We have already signed up with Trees for Travel, an organization planting trees in the rainforest, to close the gap. One tree over its life will sequester the carbon emitted in 4,000 passenger miles of commercial air travel. We expect to plant a lot of trees.

5 Resource Efficient Transportation

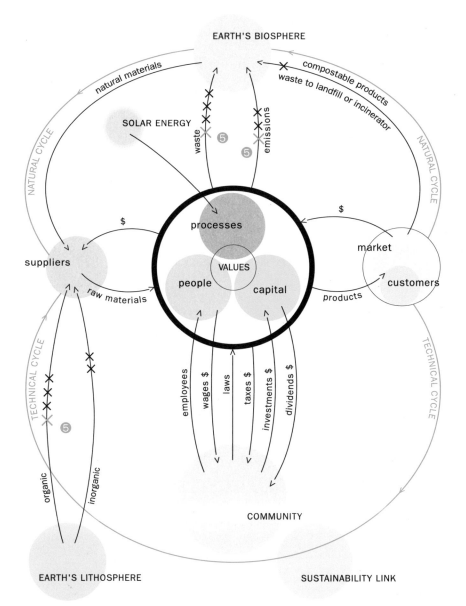

FIG. 5-10. The Sensitivity Hookup, our sixth front, spawns numerous desirable connections: service to the community through involvement and investment in the community (especially in education), closer relations among ourselves (inside the circle) to get all of us in alignment, and with suppliers and customers. (I use "sensitivity" as Brian Swimme uses it in his book, *The Universe is a Green Dragon,* meaning heightened awareness brought about by absorbing a stimulus—an influence—and being changed in the process into a new person.)

This front leads to increases among all, including our communities by way of our people, in the awareness of and sensitivity to the thousands and thousands of little things each of us can do to inch toward sustainability, breaking unwanted connections. Ties to the community, to our suppliers and customers, and within our organization are all strengthened. We hope our customers will see their role and become engaged in helping us increase our leverage with our suppliers to bring them along on the climb.

Community is redefined to include all of the community of life; our people are becoming sensitized to their stewardship responsibility for the treasure of life in all its forms, as well as Earth's life support systems. So we've adopted streams and sponsored a television program to expose the plight of our own Chattahoochee River, one of Georgia's most polluted rivers. We're planting flower and vegetable gardens on our factory grounds and creating bird sanctuaries, too.

The Natural Step becomes at once our shared framework, our compass pointing the way, and a magnet, drawing us toward the summit of that mountain that is higher than Everest, called Sustainability. The ISO 14001 environmental management system is only a threshold—a given for all our factories. It will help us track our progress.

6 Sensitivity Hookup

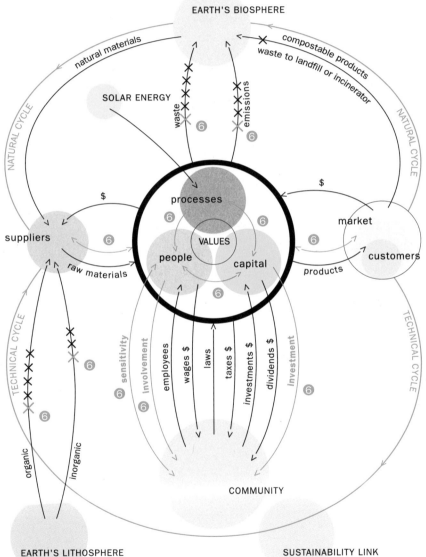

FIG. 5-11. The seventh and final front calls for the Redesign of Commerce itself. Redesigning commerce probably hinges, more than anything else, on the acceptance of entirely new notions of economics, especially prices that reflect full costs. To us, it means shifting emphasis from simply selling products to providing services; thus, our investment in downstream distribution, installation, maintenance, and recycling—all aimed at forming cradle-to-cradle relationships with customers and suppliers, relationships based on delivering, via the Evergreen Lease, the services our products provide, in lieu of the products themselves. As a result, we further break the undesirable linkages to the lithosphere and the biosphere, those that deplete or damage. Another highly desired result is increasing market share at the expense of inefficient competitors. But full cost pricing is necessary if those salvaged molecules are to be, financially, worth salvaging to replace virgin petrochemicals.

Dream a little: Maybe even the tax laws eventually will shift taxes from good things, such as income and capital (things we want to encourage), to bad things such as pollution, waste, and carbon dioxide emissions (things we want to discourage). What if perversity could once and for all be purged from the tax code? It must, for the next industrial revolution to put T (Technology) in the denominator. When the price of oil reflects its true cost, we intend to be ready. That would truly change the world as we have known it, especially the world of commerce. It's also the day we'll be kicking tail in the marketplace.

7 Redesign of Commerce

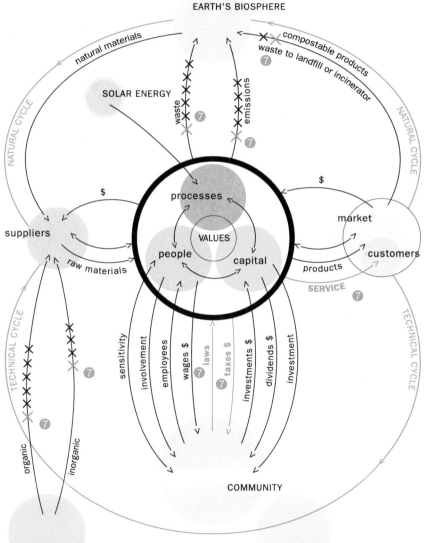

FIG. 5-12. Success on all seven fronts will bring us to our goal, The Prototypical Company of the 21st Century. What are its characteristics? It is strongly service oriented, resource-efficient, wasting nothing, solar-driven, cyclical (no longer take-make-waste linear), strongly connected to our constituencies—our communities (building social equity), our customers, and our suppliers—and to one other. Our communities are stronger and better-educated, and the most qualified people are lining up to work for Interface. Customers prefer to deal with us, and suppliers embrace our vision.

Furthermore, this 21st Century company is way ahead of the regulatory process. The regulatory process has become irrelevant. The company's values have shifted, too, and it is successfully committed to taking nothing from Earth's lithosphere that's not renewable, and doing no harm to her biosphere. The undesirable linkages are gone!

Sustainable and just, giving social equity its appropriate priority, and creating sustainable prosperity, an example for all, this company is doing well (very well) by doing good. And growing, too; it is expanding its market share at the expense of inefficient adapters, those competitors that remain committed to the old, outdated paradigm and dependent on Earth's stored natural capital when oil's price finally reflects its cost ($100 per barrel, $200 per barrel, more?). The growth is occurring while extracted throughput (materials from the mine and wellhead) is declining, always declining, eventually to reach zero. Only zero extracted throughput is sustainable over geologic time.

It makes such absolute business sense to win this way, not at Earth's expense nor at our descendants' expense, but at the expense of inefficient competitors. Most importantly, we will have proven the feasibility of moving T (Technology) from the numerator to the denominator, making technology part of the solution, and reducing environmental impact. If we can do that in a petro-intensive company such as Interface, anyone can do it. The next industrial revolution can be.

In that new era, the technophobes and the technophiles will be reconciled; the interest of labor and the interest of capital, reconciled; the interests of nature and the interests of business, reconciled. The Hegelian process of history—*thesis, antithesis, synthesis*—will lead to a sustainable society and a sustainable world. The mindset behind the industrial system will have been transformed.

Will this company lose its competitive will to win? Not on your life! The prize is now larger than ever. We must never, ever give up—not on the next heartbeat, whether it is our company's or the planet's.

Prototypical Company of the 21st Century

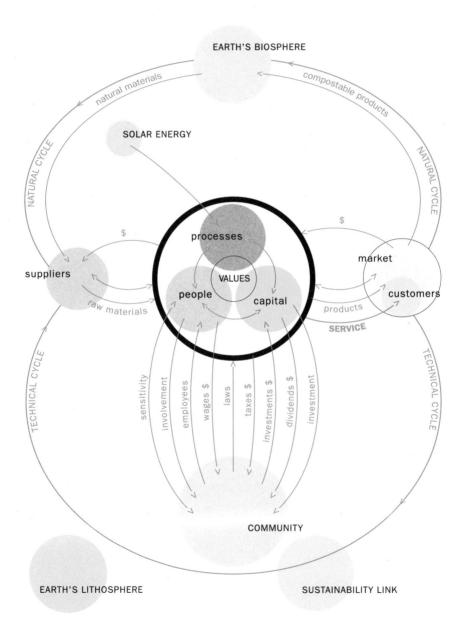

EARTH'S BIOSPHERE

natural materials

compostable products

SOLAR ENERGY

NATURAL CYCLE

NATURAL CYCLE

processes

$

$

market

suppliers

raw materials

VALUES

people

capital

products

customers

SERVICE

TECHNICAL CYCLE

TECHNICAL CYCLE

sensitivity

involvement

employees

wages $

laws

taxes $

investments $

dividends $

investment

COMMUNITY

EARTH'S LITHOSPHERE

SUSTAINABILITY LINK

The tangible results of this seven-front assault, so far? I've already mentioned our supply chain's total extraction of materials from the earth to produce 1995's sales of $802 million. That was 1.224 billion pounds of stuff from Earth's crust, mostly petro-derived. On sales of $1 billion in 1996, our second year into this journey, the total extracted material was, by our very best calculation, very nearly the same at 1.23 billion pounds, or about 19 percent more efficient usage of the same amount of extracted material as in 1995. Furthermore, we're pretty sure that qualitatively what we produced in emissions and waste streams was at least no worse than the year before. So this means that incremental sales volume of $200 million was achieved with practically no additional throughput of extracted material or harm to the biosphere, therefore at no additional cost to Earth. Several factors contributed to this: a strong shift to service through downstream distribution, waste reduction through QUEST, product redesign to require less throughput, higher prices for our products (thus expanding profit margins), and various other prod-uct mix factors. Frankly, we were surprised by the apparent progress. Yet, we believe the measurements are valid; this is the first $200 million or so of truly sustainable business that we have realized in this long climb, and it felt good!

Progress in 1997 was less dramatic, but did show further improvement in resource efficiency. With some estimating to establish our baseline year, 1994, after three years we find ourselves 22.5 percent of the way up that mountain. In three years, the pounds of material extracted from the earth and processed by our

FIG. 5-13

Progress Toward Sustainability

in extracted pounds per dollar sales
in percent of sustainability

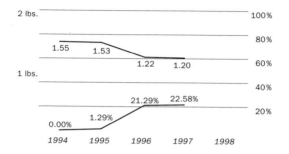

FIG. 5-14

Total vs. Sustainable Sales

total sales in millions of dollars
sustainable sales in millions of dollars

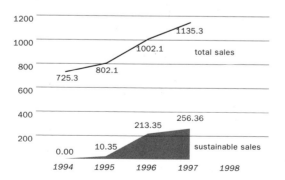

entire supply chain (including energy) has come down from 1.55 pounds per dollar of sales to 1.20 pounds per dollar. That's the 22.5 percent improvement—de-materialization—which, in 1997, yielded $256 million of sustainable sales. (See Figs. 5-13 and 5-14.) That felt good too. It's a start, but the top of the mountain is a long way away. The top means zero extracted throughput per dollar of sales and no harm to the biosphere.

Meanwhile, the new thinking—the new mindset—is beginning to permeate everything we do, especially product design and development. In our textile business, we've introduced fabrics produced from 100 percent recycled polyester and shifted entire product lines from virgin to recycled fiber. This has been accomplished, but not without painstaking and excruciating effort on the part of ourselves and our suppliers. It's hard work. Further, we've instigated the development of polyesters that use no antimony as a catalyst (leaving that poison in the lithosphere)— a significant technological breakthrough for our fiber producer/ supplier. In our carpet tile business, a recent product introduction is produced by the fusion bonded method (our initial carpet tile technology) with 72 percent closed loop recycled content—fiber-to-fiber, backing-to-backing. We call it "Dèja Vu," harkening back to our beginnings 25 years ago.

The most exciting product, however, will be introduced soon after this book is published. It will be our most nearly sustainable product yet, designed according to every principle of sustainability we know: zero manufacturing waste, even installation waste to be recycled; every raw material component is recyclable; the main production process is solar powered. The product reflects maximum de-materialization, meaning more value, less stuff. It will be offered under the improved Evergreen Lease, now called Evergreen Service Contracts. We think it may be the first product ever consciously and deliberately designed for the next industrial revolution. At the moment, we refer to it as "Project Amory." Watch for it under another name and expect a big splash in a marketplace that we believe is ready for a whole new category of flooring. That's what the new product represents—the first of a new kind. It's not perfect and we will continue to improve it, but it's a big leap forward.

Interface, Inc. made its first published progress report to the public in November 1997, the *Interface Sustainability Report*. It describes the problems of industrialism as we see them, the solutions we are pursuing, and where we are in our climb. It also makes clear how much farther we have to go. To our knowledge, it is the first such report ever produced by a corporation. It has been widely distributed to the company's stakeholders and is available to anyone else who requests one. Contact our website (http://www.ifsia.com).

Other things need to change during the next industrial revolution. New technologies and manufacturing methods, tax shifts, and products of service are not enough. By every means possible, *extractive* throughput per unit of sales (gross domestic product on the national level, gross global product on the worldwide level) must be pushed toward zero. Sustainability depends ultimately on getting all the way there, to zero extractive throughput, given the perspective of geologic time and all the time yet to be. Again, we must remove the word *extracting* from the dictionary's definition of technology.

❖ ❖ ❖ ❖ ❖

In October 1996, I was invited to Amsterdam to speak to the worldwide partners' meeting of one of the large international accounting firms. Since accountants tend to be highly analytical and unemotional, I thought that my standard speech, which has considerable emotional overtones, should be augmented if I wanted to connect with this group. So I added these comments just for the Amsterdam audience:

Now, what does this discussion have to do with you, your profession, and this meeting? I want to suggest that you and your profession are the scorekeepers in the game of business, but the rules of the game will change during the next industrial revolution; therefore, the method of scorekeeping will have to change, too, as business and commerce, and civilization, are reinvented. You could, with an early understanding of what might be, lead this change and help turn humankind from its course of self-destruction, unless, of course, you would rather just keep score as the world collapses around you. You could, for example, help to develop the field of EcoMetrics and help us understand God's currency, which certainly is not dollars or guilders, nor even pounds sterling.

I know that already you are faced with assessing environmental liabilities, but let's go further. For example, let us consider how we value *assets* today. Take a forest, a stand of trees. What is its value? I think most would say: x boardfeet of lumber at y $ per boardfoot equals z, less the cost of harvesting; that is the value.

But let me tell you a story about a small city on the banks of the Chattahoochee River in west-central Georgia in the United States, which in the first 100 years of its existence—through years of heavy rain and drought alike—never once experienced a flood. Then one year the banks of the river overflowed and $5 million of damage occurred. So the city fathers commissioned a dike to be built at a cost of $3 million, and that dike was sufficient to prevent flooding for five years. But then there was a season of especially hard rains, and the dike was breached, and the damage was $10 million this time. Therefore, the dike was rebuilt, higher this time, at a cost of $8 million, and the city was saved from flood for another seven years. And then, wouldn't you know it, the floods came higher still and the dike was breached again and someone finally said, "What is going on here?" So a team of experts was engaged to analyze the problem and one of the experts was an ecologist. And he, with brilliant insight, looked where? Not at rainfall records, nor at dike construction, nor at laminar or turbulent flows of a river. No, he looked upstream. And what did he find? He found that the forests for 50 miles upstream had been clear cut over a period of 20 years and the clear cutting had changed the hydrology of the area. Root systems no longer existed to hold the rainfall, so the rain ran off into the streams and rivers, eroding the land in the process and filling the river with silt and—by the way—killing fish, too, depriving the poor people of the area of one source of sustenance, while flooding the plains downstream, including the unfortunate small city.

So, the question arises, "What is the value of a forest?" The short-sightedness of conventional economics lies exposed, naked, does it not? And I have not mentioned the value of a tree in removing carbon dioxide, a greenhouse gas, from the atmosphere, sequestering carbon, and producing oxygen for us to breathe, nor the songs of birds that are heard no more where the forests used to be. Neither have I mentioned the disease spreading insects that now proliferate unchecked because the birds, their predators, are gone, resulting in an increase of encephalitis in the children in the region. So you see, there are serious questions to be raised about the traditional calculation of profit on the sale of the timber harvested from that clear-cut forest.

The ultimate solution to the flooding, pursued by our federal government in its dubious wisdom, was to build a dam at a cost of $100 million, which took 28,000 acres of prime agricultural land out of use and destroyed the habitat of uncounted creatures. Today the lake, thus created, is a polluted cesspool, collecting Atlanta's sewage.

The value of a forest? Think again. *(Though based on actual facts, the story is largely apocryphal and exaggerated, but I tell it for effect. I do know this river, though. As a boy I caught 20 pound channel catfish there that our family would eat for a week. Channel catfish no longer exist in the river.)*

Or, staying with assets, what is the value of a mine, say a uranium mine—something that at first blush would seem to be highly treasured? But on second thought, when we consider the cost of the nuclear clean-up that Earth faces, somewhere between $300 billion and $900 billion, depending on just how bad the Russian and Ukrainian situations turn out to be, uranium somehow seems not to be so valuable anymore. Think of the liability we have transferred to future generations! Enlightened accounting would figure out how to take that liability right into the evaluation of that mining asset today.

Let's look at Gross Domestic Product (GDP) for another exercise in new vs. old economics. Consider, for example, that the *Exxon Valdez* disaster in Prince William Sound *added* to GDP. Reflect on that. Reflect also on the absurdity of the fact that the medical expenses for a child dying of environmentally related cancer *add* to GDP. And that the costs to clean up and rebuild after a hurricane caused by global warming *add* to GDP. Clearly, as a measure of standard of living, much less as a measure of progress or well-being, GDP is sorely lacking.

I spoke earlier about the cost of a barrel of oil, compared with its price, and how the market is oblivious to the notion of external costs, both those passed on to our neighbors and those passed on to our grandchildren, what I've called intergenerational tyranny. We must think more about present value discount rates. Perhaps they should be *negative,* increasing the present value of future liabilities, rather than decreasing them. In taxation policy, the earth cries out for a carbon tax to increase the price of fossil fuels to internalize the societal costs of military power in the Middle East and global warming, and thus hasten the development of alternative energy sources.

Herman Daly, an economist at the University of Maryland in the United States, has been considered a kook by mainstream economists for years. Daly criticizes conventional economics as "empty world" economics and the economics of "unlimited resources" in what's clearly an emerging era of a "full world" with physical constraints and finite resources. Daly thinks economics must recognize reality and acknowledge that Earth's capacity to provide and endure is, in fact, limited and not infinite. People are now listening to him, even those who once derided him.

I think you should rethink economics and accounting. I urge you to think about EcoMetrics, to join the search for God's currency. Talk with Herman Daly. Change is coming. Change creates opportunity. A growing number of companies are beginning to think differently about their scorekeeping. It's just a matter of time (during the next "two-hundredths of a second") until all will have to. You could lead the way in this, and you should—for Earth's sake and for our grandchildren's sakes.

I could not tell from the immediate feedback whether I did, in fact, connect with that audience. There were a lot of stony faces; it would be unkind to describe them as blank. Afterward, there was just one request for a copy of my speech. I suppose time will tell; one never knows when a seed has taken root. The head of the firm did write me to say that my thoughts would not be ignored. I took great heart from that.

CHAPTER SIX

THE POWER OF ONE

"BRIGHTEN THE CORNER where you are." What if everybody did it?

Most of the time, when I make an environmental speech, I'm "preaching to the choir." Yet I am greatly encouraged and believe that the choir is growing, that the global brain *is* waking up. The number of "alarm clocks" to wake us is growing, too.

A quotation attributed to President Lyndon Johnson comes to mind. A rancher and landowner in his native Texas, Mr. Johnson was asked just how much land he wanted to own. He was said to have answered, "Well, just what I have and all that's next to it." Though Johnson's quote illustrated pure greed, in a filial way that's the choir we want singing the gospel of change: what's there now and all that's next to it. So, to this swelling number, I continue to say that we are all part of the continuum of humanity and the web of life in general. We will have lived our brief span and either helped or hurt that continuum, that web, and the earth that sustains all life. It's that simple. Which will it be? It's your call.

How can we help? I believe one person can make a difference. You can. I can. People coming together in organizations like yours and mine can make a big difference. Companies coming together, for example, customers and suppliers uniting in recycling efforts, can make a vast difference. Harnessing wind, current solar income, and hydrogen can make a monumental difference. Daniel Quinn's mission in his paradigm-shifting novel, *Ishmael*, is to change the minds of six billion people. If that happened and they decided to live their daily lives with Earth's welfare in mind, then Earth, humanity, and all the continuum of life would, indeed, gain a new lease on life. The mid-course correction I think Earth and humanity need probably depends on, more than any other one thing, changed minds, i.e., new paradigms. I have suggested one for business: Doing well by doing good. But what will power this change?

In June 1996, Interface sponsored an event called The Power of One. We had seven speakers in the course of a day and a half together: David Brower, Bernadette Cozart, David Gottfried, Paul Hawken, Emily Miggins, John Picard, and Johnna Wenburg.

Each told us about his or her work. There was a wide variety: Two authors, Paul Hawken and David Brower, Hawken, Chairman of The Natural Step USA and the philosophical heir-apparent to Brower, an 84-year-old ex-mountain climber with a reverence for nature who had been the country's most influential environmentalist for 50 years, dating from his days as Executive Director of the Sierra Club.

Bernadette Cozart, greening Harlem and instilling beauty in a community, along with pride and self-respect in its people. Emily Miggins, saving trees and speaking up for women's rights in India and China, as well as right here at home. Johnna Wenburg, telling about living with the orangutans in Borneo and keeping a chimpanzee orphanage in Africa. John Picard, making a difference in construction projects and in architectural and interior design circles all over this country. And David Gottfried, giving up the real estate development and investment banking businesses to follow his heart and live out his love for Earth by promoting "green buildings."

The Power of One has become a recurring theme in our company, as many of our customers, as well as our people, recognize. That day and a half set me to thinking about the power of influence and the way all of those people were exerting influence that went far beyond the direct, immediate effects of what they were doing. Though we, as a company, have a very long way to go to sustainability, how far notwithstanding, I realize now that the journey is taking place on three levels: 1) the level of *understanding* sustainability, 2) the level of *achieving* sustainability, and 3) the level of *influence*. All this can be illustrated graphically; I use Interface merely as an example (See Fig. 6-1).

The curve of *understanding* or knowledge, call it the "learning curve," is not only about learning what and "where" sustainability is, but also about how to get there, including identification of the technologies, attitudes, and practices that are needed and how they

FIG. 6-1

The Sustainability Curve

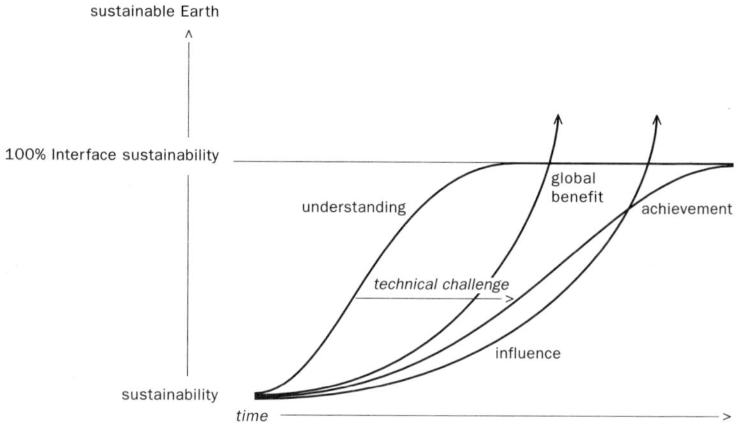

should be developed. Getting well up this curve has brought us to the seven-front plan for climbing that mountain called Sustainability, as well as to the model for the sustainable enterprise.

The curve of *achievement*, call it the "doing" curve, plots the substantive progress toward sustainability. The gap between knowing and doing represents the *technical challenge* (just not knowing how) or possibly the resource gap (not being able to afford it), but not, we hope, a gap of commitment nor of will power. In both understanding and achievement we can only hope to approach 100 percent sustainability asymptotically; 100 percent is the limit. Getting well up the curve of understanding first, truly *getting it*, is imperative if the achieving part is to be intelligently directed; it's easy to take wrong turns and tangents such as down-cycling precious molecules into less valuable forms—to get it wrong, even with the best of intentions. There are too many right things to do and time is too short to waste resources on tangents. But understanding without those actions that lead to achievement is feel-good hypocrisy and mental self-gratification.

The third curve of *influence* is the one that will take our company beyond sustainable to restorative, putting back more than we ourselves take, and doing good to Earth, not just no harm. The benefit to Earth from inspiring others to take action too can be greater than from the company's own achievement. On the front of influence, it's worthwhile recognizing that, while there are limits to power from most sources, the power of influence has no limits (as my friend Dr. Zink reminds me regularly). There are no asymptotes, no mathematical limits, for influence. Yet influence collapses without the undergirding credibility that comes from actually doing it.

The fourth curve, the resultant curve that combines influence and achievement, may be labeled the *global benefit* curve. A company's total contribution is the sum of what it does and what it helps, inspires, or otherwise influences others to do. Like influence, global benefit also may be limitless; it knows no asymptotes, either.

Finally, it's worth pointing out that this same graphic representation can apply to an individual person as well as a corporate community of people. In fact, the graphic picture of an organization in many ways is the sum total of all its individuals' graphs.

Each of us individually is one in 5.8 billion (at this writing, but growing); yet all of us at Interface are 7,000 in 5.8 billion, more than one in a million. But Interface is one thirty-three-thousandth (1/33,000th) of the Global Economy. DuPont and BASF, both 50 times bigger, our major suppliers, are each one six-hundredth

(1/600th)! All of us are resource intensive, so together we have an even bigger impact than our numbers suggest. As a company, Interface can make an immense difference by setting an example, especially if we can influence DuPont and BASF to join our efforts—and Solutia (spun off from Monsanto), and Geon, and Shell, and all the rest of our suppliers. If our customers join in, too, we can make a colossal difference!

You (the reader), too, have influence. You have the power of one. Your organization has influence, too—the collective influence of one and one and one. Knowledge, deep (not superficial) knowledge, getting well up that curve, comes first. Doing (taking action) must follow—in your personal lives and at work. Knowledge and action are critical. They give credibility and validity to your examples and to your influence, which can spread and grow without limit. You, too, can join in that positive feedback loop, doing well by doing good, a win-win for you and Earth.

❖ ❖ ❖ ❖ ❖

I've loved Georgia Tech as long as I can remember. My fondest memory of childhood is sitting on my back steps on a crisp November afternoon, eating a pomegranate that had burst open from an early frost, and listening to the Georgia Tech football game on the radio. I remember Dinky Bowen kicking a field goal to beat Navy 17-14 in 1944. I remember Johnny McIntosh tackling the Navy fullback at the Tech goal line in 1945, forcing a fumble that a freshman named George Mathews picked out of the air to run 95 yards for a Tech touchdown. I also remember sitting on those same back steps a few years later on a Friday afternoon, polishing my football shoes before that night's game at West Point High School, talking with Lew Woodruff, an assistant to Coach Bobby Dodd, as he recruited me to come to Tech.

When the grant-in-aid (football scholarship) offers came from Auburn (Coach Shug Jordan), Georgia (Coach Wally Butts), Kentucky (Coach Bear Bryant), and from Tech (Coach Dodd), there was never any question in my mind about where I would go. That shoulder injury my sophomore year ended my football career, and I was able to concentrate on my studies, work hard, do well, and graduate with my degree in Industrial Engineering.

Thirty-three years later, well into a career as Founder and CEO of Interface, Inc., I found myself on the Georgia Tech Advisory Board (GTAB), a sounding board for the president of the institution on strategic issues. After six more years, I became Chairman of that board—at a most propitious time. It was the first year of new President Wayne Clough's tenure.

In his new role, Dr. Clough set out to establish his own agenda for the institution, part of which was to engage the entire faculty and administration in a collegial effort to rethink and rewrite Georgia Tech's vision and mission statement. In due course, a first draft materialized and, as Chairman of GTAB, I received a copy.

To my dismay, the notion of sustainability could not be found in the statement, nor the words *environment* or *ecology*, even though Tech, with funding from the General Electric Foundation, had had a Center for Sustainable Technology for more than a year. (Talk about the power of one and being in the right place at the right time!) Suffice it to say that the final version, the published version, commits Tech to work for a "sustainable society" in the lead paragraph, and has other references to sustainability and the environment throughout. Sustainable Technology has joined Biotechnology and Telecommunications Technology as one of the three major thrusts, or strategic areas of study, the university is mounting for the 21st Century.

I am also involved with my old school, now renamed the Industrial and Systems Engineering (ISyE) School and ranked number one in the nation by *U.S. News & World Report* for six consecutive years. As it would happen, we have a capital campaign underway, along with the rest of Georgia Tech, and I am Chairman of the ISyE Capital Campaign Executive Committee. Elsewhere in this book I have talked about the kinder, gentler technologies of the future which I believe must emulate nature. Well, I've put my money where my mouth is, and committed Interface and myself to endowing a chair within the ISyE School, the Anderson-Interface Chair of Natural Systems. I don't know of another such chair in any university. Dr. John Jarvis, the head of the ISyE School, has a standing challenge from me: to learn how a forest works. It's the work of a lifetime, but I believe as the forest's symbiotic relationships are understood, new organizing principles for industry will be revealed. Those, I believe, will shape the model for industrial systems in the 21st Century and the next industrial era, and will define the means by which Interface, as the prototypical, sustainable company of the 21st Century, will operate. I think Georgia Tech can and should lead in discovering this model. The timing is perfect. As Tech is planning a transition from the quarter system to the semester system, the curriculum must change, too. Biology will become important to engineers. That's an amazing change!

In a sweet irony, the friend who challenged my views and sent me the books that presented "the other side" has endowed a chair in another department of Georgia Tech: The Chair of Environmental Biology. The distinguished professors who occupy those two chairs are going to have a lot of fun together, and Georgia Tech's strategic focus in the 21st Century, new curriculum and all, will be profoundly influenced.

It also pleases me enormously that Kalle Robèrt, having been introduced to the Georgia Tech community through my efforts and after brilliantly presenting the principles of The Natural Step to a campus gathering of some 200 scientists and engineers, has experienced a surpassing acceptance at the institution by being invited to, and accepting, a position on the Dean's Advisory Board of the College of Science. The Dean, Dr. Gary Schuster, was the self-pronounced chief skeptic of The Natural Step until he had had an hour long, one-on-one discussion with Kalle, a session from which I watched them walk away arm in arm after the most intense of discussions about the sustainability (or unsustainability) of nuclear power. I had the enviable privilege of being the proverbial fly on the wall. It was wonderfully enlightening! (Their conclusion about nuclear power? Yeah, maybe, IF nothing went wrong. Who wants to risk it?)

One result of Kalle's influence is that Georgia Tech's Center for Sustainable Technology has embraced The Natural Step and its principles of sustainability. Another result is the creation of an interdisciplinary Sustainability Task Force to augment the efforts of the Center for Sustainable Technology in sensitizing the entire campus. At one level, this task force's mandate extends to influencing daily life on campus and, at another level, to instilling the teaching of sustainability so it becomes ingrained in every course, and at still another level, to research and outreach, to further the embracing of sustainability throughout society, especially Georgia-based industry. Good people are involved. Good things will result. Tech's role will continue to expand. Of this I am sure.

I look for Georgia Tech to be a force in America's climb toward sustainability, and the Center for Sustainable Technology, embracing The Natural Step, to be a powerful catalyst in generating that force.

CHAPTER SEVEN

TO LOVE ALL THE CHILDREN

DR. GARY SCHUSTER *of* Georgia Tech *is the source of a most provocative story, which he says is true. He says that the scientists who developed chlorofluorocarbons (CFCs) to be refrigerants, propellants, and stain-resistant coatings knew they had invented inert compounds. They thought that the compounds, because they would not react with other elements, might last forever. And they knew the compounds would accumulate in the stratosphere. So they could see CFCs lasting forever, accumulating and accumulating in the stratosphere. Could such compounds be produced and marketed in good conscience? They actually raised the question!*

But someone reasoned that no, the compounds, once they were in the stratosphere, would be attacked by ultraviolet radiation and break down into their basic elements, chlorine, fluorine, and carbon; so, no, they would not last forever.

Unfortunately, no one asked the most important question of all, "And then what?" And then what, we all learned many millions of tons later, was that the free chlorine in the stratosphere would attack the stratospheric ozone layer, resulting in the serious weakening of the ozone shield, which protects life on Earth from deadly ultraviolet radiation. Eventually, the unasked question was answered at great cost, not only to the producers of CFCs, but to society as a whole.

"And then what?" is an important question.

So, and then what? Where does it go from here for Interface? Suppose Interface is very successful in its efforts toward sustainability. Suppose, further (for example), that global warming, to cite just one threat, turns out to be so vividly demonstrable and undeniably true that the whole world wakes up one day with a gigantic cry of alarm. Suppose consumer outrage erupts and markets shift overnight. Where will the capacity to respond be found? A lack of capacity for response could be a stupendous stumbling block to Earth's welfare.

Without early and gradual tax shifts or tradable emission credits to effectively tax carbon and to allow markets to adjust in an orderly way and bring new technologies on stream, a repeat of the oil shock of the 1970s is highly likely, should the unhappy wake-up call come to an unprepared world.

At Interface, we see that alarming possibility as an opportunity for which those with foresight should prepare. Thus, Interface has created One World Learning (OWL), a company within our company, whose mission is to be the repository for the wisdom and knowledge we are accumulating on this mountain-climbing expedition and to impart it to others. OWL's genesis can be traced to a brainstorming session to conceptualize a new, sustainable enterprise. We told a task force, "If you can dream up a sustainable business, we'll create it." In selecting OWL's name, we have avoided using the word "teach" in recognition of the implicit wisdom of the Iroquois language which (I've been told) has 35 words for "learn," but not one single word for "teach." OWL is in the *learning assistance* business.

OWL is sharpening its skills internally, first, by helping our people learn The Natural Step, using the pedagogy developed by Karl-Henrik Robèrt. OWL will help other companies, too, to learn The Natural Step, and then move beyond The Natural Step to other areas we are pioneering. For example, experiential learning, team building, personal breakthrough, and value shifting are facilitated by OWL under "Why?"—the program name we have adopted. We want Interface always to be a learning organization. OWL is integral to that commitment.

For another example, as a company we know a lot about merging cultures. As a river takes on the waters of its tributaries to grow wider and deeper, Interface has grown with some 48 acquisitions since 1982. There are many valuable lessons we have learned from those acquisitions and the companies we have acquired as they have been absorbed and hooked up. Further, the same methods for facilitating cultural synthesis have helped us open up our people, via the Why? experience, to the personal breakthrough changes that QUEST and EcoSense have needed our people to make, to become an empowered team. Command and control management would not have gotten the same results.

QUEST and EcoSense were part of a larger reinvention of Interface, Inc. No one and no company are born knowing these things. We hope to be in a position to help others who want to follow to short-cut the learning process and soften the licks from the school of hard knocks. God knows we have taken a few along the way. And had some major triumphs, too—like Maui.

Let me tell you about Maui. April 6, 1997 marked the twenty-fourth birthday of Interface, and the beginning of our twenty-fifth year. A year or so before, I had casually mentioned to Charlie Eitel, then soon-to-be-President and Chief Operating Officer of Interface, that I had thought someday we might have a worldwide meeting of all of our different subsidiary companies' sales forces, in one place, together. As our management team had cobbled Interface together over the years with those 48 acquisitions, we had left each company to conduct its own sales meetings—normally once a year in the winter months early in the new year.

With Charlie, who is naturally aggressive and who also has been exposed to the high rollers of the Young Presidents' Organization (YPO) (something I barely missed becoming a member of by not leading Interface to reach $1 million in annual sales before

I reached age 40), you mention something like "worldwide sales meeting" but once. Before I knew it, Charlie had rented the Grand Wailea Hotel and Spa in Maui, Hawaii for the entire week in which April 6, 1997 would fall. Why wait until the twenty-fifth anniversary? Why not kick off the twenty-fifth year with a big twenty-fourth birthday party, and celebrate all year long? Charlie knew the Grand Wailea from a YPO meeting. I had never seen it. But we had bought it for a week and had to figure out what to do with it. With uncanny foresight, Charlie began as soon as he had bought the week to urge the hotel management to think in environmentally sensitive ways—about recycling, for instance, a foretaste of what was to come.

In due course, with much planning, a theme emerged, "One World, One Family, A Celebration" and an objective for the meeting became clear: to take this far-flung company, so painstakingly assembled over the years to fulfill a grand design, and hook it up. A corporate strategy, *diversify and integrate worldwide,* had seen *diversify* and *worldwide* executed much more effectively than *integrate.* Our meeting in Maui could complete that strategy and generate a shared sense of values at the core to provide the centripetal, binding force to coalesce this cobbled-together company into a unified whole.

Some sub-themes emerged, too, the kinds of themes that, if developed properly during the week, could facilitate the realization of the primary objective: to hook it up. Among the sub-themes were *people, product, and place,* the three Ps on which we had focused as a company in the larger reinvention process that had started in

1993. I undertook to lead the development of *place,* the environmental thread, one of the three we would hope to weave into the fabric of the meeting throughout the week.

A chance to make a stepping stone out of a stumbling block immediately presented itself. We also had reserved the Lake Placid Lodge in upstate New York for a three-day conference that had been planned for customers. But the sign-up had been so meager that we had decided to cancel the conference. With the hotel paid for and no conference to use it, we decided to salvage our cost and use the lodge for a planning retreat for the place team, which included our environmental consultants, the group I had dubbed the EcoDreamTeam.

At that time, September 1996, the DreamTeam consisted of John Picard, Paul Hawken, Daniel Quinn, David Brower, Bill McDonough, and Amory Lovins. This newly formed team (with John Knox filling in for David Brower, who couldn't be there) went to Lake Placid. Once there, the team learned for the first time where the meeting was to be, The Grand Wailea Hotel and Spa on the island of Maui.

There was instant consternation, then revolt! Hawken's face clouded, and he asked, "Why there? That's the most extravagant, expensive, opulent hotel ever built. It would be totally inconsistent with Interface's efforts for the environment." In so many words, "Count me out!"

Quinn echoed the sentiment—others joined in the chorus of protest.

Totally taken aback, never having seen the place, but thinking of the sunk cost, I stood my ground. "We are going," I said, "with or without an environmental theme."

Searching for a middle ground, someone said, "Maybe we could do something environmentally sensitive at the hotel."

Paul said, "That misses the point; the bigger issue is how tourism is destroying the islands and the native culture."

I asked Paul whether he might be able to get a native-born speaker to address our meeting and enlighten us.

He said, "Yes, but you know what they'll say, don't you?"

"No," I puzzled.

His stunning reply brought our tense discussion into perfect focus, "If you asked the native people of Maui, the indigenous Hawaiians, 'What can we do for you?' they would answer, 'Don't come.'"

It was a standoff, a tense one at that. But Bill McDonough saved the day. "Why don't we," he offered, "treat The Grand Wailea as a metaphor for how we live our lives? Let's make it a design problem. If we can change the hotel for the better, and use it as a classroom in the process, maybe we can all learn how to live our personal lives better." The Power of One, and what if everybody did it? Perfect! It was a brilliant idea. It saved the day and more. All the team members came on board and the ideas began to fly. The other linchpin idea also came from Bill: "Let's do something while we're there to love all the children."

The planning meeting was a great success. Now came the challenge of executing. Would the hotel cooperate? Would they allow us to change anything? What would we change if they would let us?

To get at all these issues, we organized a visit to Maui in November 1996. I used October to arrange to meet Takeshi Sechiguchi, the Japanese owner of the hotel, approaching the introduction in the Japanese way, through a Japanese friend, Mako Yasuda.

We hoped, through the efforts of our own Jim Hartzfeld, together with John Picard, Paul Hawken, and Bill Browning from Amory's organization, The Rocky Mountain Institute, to eco-audit the hotel and gain some idea of the changes our people, as guests, could make to lessen our collective environmental impact, or footprint, while we were there in April. We hoped further that the hotel staff might be inspired, too (the influence curve). But we thought the chances pretty slim without Mr. Sechiguchi's support.

Our first day there in November, Charlie Eitel and I sat down with Sechiguchi-*san*, his interpreter, and his American General Manager, Greg Koestering. Immediately, Sechiguchi-*san* understood our objective and, apparently, the opportunity. As soon as we told him what we wanted to do and suggested that his staff could learn with us in the "classroom," he turned to Greg and, without the interpreter's help, said, in perfect English, "Whatever they want, do it."

Wow! All the barriers came down; the caution vanished; the hesitancy evaporated. By the next day, the word had spread throughout the hotel staff. The day after, it was all over the island and demand began to build to let the islanders in on this happening, too.

The eco-audit proceeded swimmingly. The hotel, we learned,

had comprehensive measuring systems for resource usage: electricity, water, propane, solid waste, detergent usage. The staff knew some of the problems, too: suntan lotion in the swimming pool, for example, clogged the pool's filter system and increased electrical loads on the pumps.

The Grand Wailea was, to be sure, an energy hog, but it did provide an absolutely superb guest experience (thus its appeal to the YPO). In planning our changes with the hotel staff, we undertook not to institute changes that would diminish the guest experience.

With the eco-audit done and the impact-reducing ideas in hand, the planning accelerated as April approached. The number of attendees was coming into focus by now: 1,100, including 100 people to be sent by our suppliers, almost 100 support or resource personnel, and some 900 Interface associates from around the world (among them, 50 people from our factory floors, chosen in random drawings at each plant to represent their associates who were left at home keeping the plants running). But how to communicate with this international assemblage representing management, sales and marketing, and factory associates from 34 countries and speaking eight languages? How to sensitize them, get them on the same wavelength?

Hawken, Browning, and Hartzfeld set out to design an exercise to begin to do just that. They called it "The Global Village." In a group the size of ours, one person could represent about five million people of Earth. What an opportunity! To put 1,100 people in a room, representing the whole Earth, and illustrate the distribution of population, the maldistribution of resources, and misery in its numerous forms: hunger, disease, ignorance, want, mistreatment of women. We wanted a mind-shifting experience.

The crowd began to assemble in Maui on April 4, 1997. By April 5 everyone was there. A Polaroid photograph of each person was taken on arrival for a special purpose to be announced later. The first evening was an opening hospitality session—beautiful, smartly dressed people, buffet dinner on the hotel's luxurious grounds overlooking the Pacific—just what everyone would expect in such an opulent setting. A hint of things to come was offered as Jennifer Eitel, Charlie's daughter, led the Maui Children's Choir in concert for us. As they sang, "We are the world, we are the children," throats constricted a bit in the presence of such precious innocence. Still, it was your normal, glorious opening evening, but after it was over we collected and weighed all the wasted food— information for later.

The next morning, April 6th, all 1,100 participants assembled. The sheer magnitude of the undertaking to bring the group together brought goosebumps. Charlie recited one by one the names of all 34 countries where Interface people are located. Representatives from each country stood. The applause was enthusiastic, with the smallest contingents receiving the loudest welcome—demonstrating the spirit of support developed among our associates on the ropes courses of One World Learning and its forerunner, Pecos River Learning Center. Hook-it-up had begun there.

In my own opening remarks, I shared a startling fact, ascertained by Dianne Dillon-Ridgely, one of our directors: It was exactly 1,000 days until the millennium. The stage was set.

The balance of the morning brought a series of inspirational speeches, highlighted by Terry Waite's jarring personal account of his imprisonment at the hands of his Lebanese captors for 1,763 days, most of the time blindfolded and in solitary confinement. Terry, a giant of a man physically, was powerful and moving, a testimony to the strength of a giant human spirit, too, standing on truth in the face of great adversity—the *people* theme of the meeting. And he provided the perfect, sobering preparation for what was to follow, the sensitizing session of the Global Village. Blind balladeer Ken Medema concluded the opening morning with a song improvised on the spot and just for the occasion. He summarized the extraordinary spirit that had already taken hold with, "We'll Get it Right This Time"—a theme song for the next industrial revolution.

That afternoon, everyone took a seat in a U-shaped arrangement of chairs. In every seat had been placed a piece of paper containing some numbers. Hawken was master of ceremonies. He introduced the purpose, to give us all a better understanding of who we are, and the tragedy, the sorrow, as well as the celebration to be found in the human condition. We were instructed to stand as the numbers on the papers each of us had taken from our seats were called.

Bill Browning then described Spaceship Earth, a speck in the cosmos moving 800 times faster than a speeding bullet. He acquainted us with the scale of the solar system and the minuteness of Earth, especially what we know as its extreme contours, Everest and the Marianas Trench, represented by the thickness of a dime and a nickel, respectively, on a dymaxion projection of the Earth laid out on the floor, 1:6,000,000 scale. Humble speck we are, indeed.

After I presented David Brower's "Six Days of Biblical Creation" to fix in people's minds the geologic time scale (some hearing it for the first time), Paul proceeded to the exercise, beginning with a film that graphically demonstrated the population explosion from the beginning of the Common Era (A.D. 1) to present, and projected to the year 2020. It left everybody gasping, "My God!" as a map of Earth literally filled up with lights, each light representing a million people.

Various commentators had been selected to help orchestrate the exercise. The first request from a commentator came, "Would all who have the number one on your paper please stand."

The entire room stood. "You are the people of Earth. Please be seated."

Paul explained that we were simulating Earth's human population, one per five million.

The next commentator, "Would all who hold the number two stand."

Number two represented the 486 million people of Latin America; 93 people stood and were asked to remain standing. Then the United States and Canada: 56 people stood, representing 295 million, and remained standing. In like manner we progressed around the room to Russia and Eastern Europe (59 people, representing 309 million), Africa (140 people, representing 732 million), the Middle East (34 people, representing 176 million),

Europe (80 people, representing 419 million), Indonesia (38 people, representing 201 million).

Then we came to Asia. A breathtaking crescendo began to build. Japan (24 people representing 126 million), Asia less Indonesia, Japan, China, and India (158 people, representing 830 million); then India, and 181 people stood, representing 950 million people! Finally China (everyone could see it coming), 232 people stood for 1 billion, 218 million. Whoosh! The collective gasp was audible, and nervous laughter punctuated the buzz. More than half the room represented Asia!

As the audience settled back in their seats, a commentator called the next number and 26 people stood, representing the 135 million babies to join humankind this year. Then, another number, and ten people sat down, representing nearly 50 million who would leave us by dying. That left a net gain in population of 85 million, with 16 people still standing.

Paul returned for the next revelation, looking ahead to the future. He set the stage again by referring to the population film shown earlier. "It took all of human history until the year 1800 for us to reach a population of one billion. One hundred years later we were two billion. About 50 years later, we reached three billion, and 25 years later in 1975, we reached four billion. In 1985, we reached five billion. Today, 1997, we are at 5,771,000,000. The rate of population growth has slowed in the last few years; however, at current rates, how many more will there be in 50 years? Would number 17 please stand." All 1,100 people stood again. A doubling to 12 billion in just 50 years!

Further looks at the future: When number 18 was called, 350 people stood, representing all the children with us today. We were reminded that *they are* the future. Then 70 stood, representing the elders over 65 years of age, who will be leaving soon.

Next we looked at the makeup of our population today, at what we do for a living: five stood, our soldiers. Seven stood, our teachers. One stood, our village's doctor. Eight stood, our 40 million refugees. Our farmers stood next, 143 strong, 13 percent of our population, down from 40 percent in 1950. *All* workers, the 2.5 billion with jobs, were represented by 477 villagers standing. But more staggering was the response to the next number, those *without* jobs, looking for work and unable to find it: 190 villagers stood; almost one billion human beings, victims of population growth, uneven economic development, and labor saving technologies. One could not help but think of the 16 additional villagers (85 million people) who join us *every year,* most being born into those areas where most of the one billion jobless already live.

The intensity grew as we began to examine the distribution of resources and industrial throughput: aluminum, iron, steel, chemicals, paper, timber—so disproportionate to the distribution of population. Income distribution, too, was shocking: 248 stood to represent those who earn less than $370 a year! At number 28, the richest one-fifth *and* the poorest one-fifth stood together, 440 in all. We learned that the richest 220 have 61 times the average means of the poorest 220. When the billionaires were asked to stand, no one stood. Paul had a tiny, two-ounce figurine in his hand at the podium. It represented, in proportion, he said, the 300 billionaires on Earth who collectively receive as much income as three billion average earthlings.

At number 30, we began to examine the distribution of energy usage, then pollution, then food distribution. At number 33, those who go to bed hungry each night stood—187, representing 981 million! Then, 119 stood to represent the 595 million underweight and malnourished children—one in every three!

And so it went, grim fact after grim fact revealed. The plight of Earth, the vast inequities, and the vulnerability of the human species stood starkly exposed, all around. I found myself standing in the category of women in the world who have no access to birth control. I represented five million women, and I was joined by 183 (915 million) others. When eight people stood to represent the 40 million of us who die from hunger or hunger-related illness each year, Paul reminded us of the simple arithmetic that equates this to 300 jumbo jets crashing *every day!*

There were messages of hope, too. We learned that half the children in the world are now immunized against measles and polio, that life expectancy has increased by a third since 1960, that access to safe drinking water has almost doubled, that primary education enrollment is up nearly two-thirds since 1960, that food production is up in the last decade, that educational enrollment of girls has doubled over the last two decades, that infant mortality rates have halved since 1960, that three-fourths of Earth's people live in democratic regimes, that micro-lending institutions will soon serve 100 million of the poorest of us, and that the usage of energy per unit of GDP is declining rapidly. Paul wrapped it up with

Nelson Mandela's thoughts about individual responsibility and the power each of us has to make changes happen in our world— Mandela's personal and unique version of the Power of One.

The Global Village experience was unsettling as well as inspiring. We didn't want to leave people in the grips of depression or despair. Dr. Zink had suggested a period of decompression and venting, so we concluded with small breakout groups. People talked with each other about their feelings for a while; then Dr. Zink, Larry Wilson of Pecos Learning Center, and Archie Tew of One World Learning engaged the breakout groups to elicit comments. The comments came, thoughtful, articulate, representing the diversity of viewpoints in the groups. The comments of an African-American male associate from Georgia epitomized the dominant impression of the day: "I had no idea before how my lifestyle impacted women and children I don't even know in faraway parts of the world. Now I know."

Everyone was engaged and, at the end, everyone was emotionally primed for the challenge. I told them about the Lake Placid experience and Hawken's comment, "...don't come." And I expressed my wish that we be such a different group of visitors that we would be invited back gladly by the native people of Maui.

The challenge was issued: Do something to love all the children. Right there, 40 teams of 25 or so people were formed. Native Hawaiians had been recruited in advance to become a part of each team, one per team. Each team was challenged to develop an idea of what we could leave behind as a permanent legacy for the children of Maui, an idea that would survive not just our leaving the island, but a legacy that would still be there serving the children when we had all left this life. The natives were there to help the challenge teams make sure the ideas were relevant.

I then announced Interface's commitment to provide $50,000 to fund the implementation of the three best ideas, and the teams were asked to report their ideas three days hence. The teams gathered and went to work, but not before the Polaroid photographs from arrival day were explained. An entire wall of the hotel had been designated as a personal legacy wall. The photographs were displayed on the wall, and each person was invited to post personal commitments beside his or her picture throughout the week, indicating changes in behavior. At the end of the week the wall was full! More than 700 commitments to do those "thousands of little things" had been posted. The Power of One, a recurring theme the entire week, was manifest. Every commitment was preserved in the newsletter that was subsequently published to report on the meeting to all our associates left behind to tend shop and as a reminder for those who were there.

The next morning, a representative of DuPont, one of the suppliers present, without prompting by anyone from Interface, asked to speak. He announced that the supplier group would match Interface's $50,000! As the teams began to report in, there were scattered, individual offers to contribute sums of money, $20, $50, even $100. So Paul tried an idea out on the group in plenary session. How many would give $20? A few hesitant hands went up, then quickly came down. How many, $50? No hands were raised! How many, $100? The whole room exploded, every hand up! In an instant, the funding had mushroomed to $200,000 as every person there pledged $100 toward the legacy. The next day, one team announced that it wished to withdraw from the competition for a share of the funds, and just sponsor its own project, adding to the total funding the cost of their project to be borne by themselves separately. Patagonia, a supplier of commemorative shirts for the meeting, pledged its profits on the shirts.

That same morning, we announced the tactical objective of the meeting: to reduce our impact on the hotel every day during the time left, to make the hotel a classroom in conservation and sustainability. We gave the audience the consumption measurements from the first day: electricity, water, propane, solid waste (including that left from the buffet dinner on the grounds). Then, for the days left, ideas were suggested: hang your towel and use it again tomorrow as you would at home; turn out all the lights when you leave the room; take shorter showers; use the glucose-based soap and shampoo that had been procured just for us (the fish could eat it!); sleep with your windows open and let the air conditioner rest; draw the blinds during the day to keep the sun out of the room when you're not there; take only what you really need to eat.

There had been a group of nearly identical size from another American company at the hotel the same week of 1996—a year earlier. That provided a perfect baseline with which to compare resource usage. Each morning, Jim Hartzfeld of Interface Research Corporation and Bill Browning of the Rocky Mountain Institute reported on the comparisons for the day before, resource by resource. Usages began to decline as people pitched in. Even though the hotel staff had been thoroughly briefed and treated to lunch one day by Interface to "seal the deal," institutional inertia exerted its grip right away. Our people complained that the housekeepers wouldn't leave the used towels on the rack. The staff members were well trained, so change was difficult. They couldn't "untrain" themselves for this strange crowd. By the third day, though, they got it, and the resource usages declined more rapidly. One day water usage inexplicably spiked upward. What? Collective chagrin reigned! The entire group was so sensitized by now that a massive hunt for the leak ensued. No leak could be found. Devastation! Had our usage actually increased? As it turned out, the data had been incorrectly recorded. Relief! Whew! Water usage actually had gone down.

By the end of the week the resource usage stood at:

Water:	Down 48 percent, equivalent to the entire rainfall for a year on the hotel's 42 acre grounds.
Electricity:	Down 21 percent, carbon dioxide emissions reduced the equivalent of 42 acres of forest—trees freed to remove someone else's CO_2.
Propane:	Down 48 percent, equivalent to another 13 acres of forest; petrochemicals left for others.

Solid waste generated: Down 34 percent, the landfill
spared and $40,000 (annualized)
in tipping fees avoided!

Overall, financial savings were $1,081,000 on an annualized basis! Pretty good for the hotel, but Earth benefited even more. The Power of One, when everybody does it: lesson learned! (There was a nice *quid pro quo,* too, when we received a $1 million carpet order from the hotel. How's that for doing well by doing good?)

The people of the laundry, with half their regular jobs to perform, planted an experimental garden on the grounds, using native species—under Bernadette Cozart's direction (who better?) and with the help of Interface Research Corporation's people. The idea: to figure out a better, more sustainable horticultural plan for the grounds and their semi-arid setting.

The sugar-based soap and shampoo were great; the sun block provided, not so great. A lot of sunburn suggested a lot of determined, committed people who were going all out for Earth, which proved to be a bit too much for tender skin. They got lots of sympathy.

Throughout the week, one emotionally moving speaker followed another, with objective, unemotional presentations interspersed. Tom Crum, author and teacher, conducted a session on conflict resolution using aikido, a martial art in which he is expert, to illustrate and drive home his points—further weaving the *people* strand into the meeting. Paul Saffo spoke on the information technology revolution; Amory and Hunter Lovins, on energy efficiency. The EcoDreamTeam inspired everyone: Hawken, McDonough, Brower, Picard, the Lovinses, Browning, Cozart. Daniel Quinn, facing a deadline on his newest book, sent wife Rennie in his stead. Ken Kragen, retained to help visualize the plan for Maui week, inspired us with his account of organizing Hands Across America and We Are the World. The Hale-Bopp comet punctuated each day's end, as if it were there just for us, along with our very own lunar eclipse. Just offshore, humpback whales breached and spouted in seeming approval.

There was world class entertainment, too: John Denver, The Pointer Sisters, Kenny Loggins. John Denver composed and sang a new song just for us (perhaps his last original work), "Blue Water World." He obviously loved being there, and pronounced the week a turning point for sustainability. (We hope to cooperate with John's Windstar Foundation to use the tape of the Maui premiere of "Blue Water World" to honor John's memory.)

In the end, even such world class entertainment, while extraordinary, became secondary. The environmental theme completely eclipsed everything else in its effect on our people. Without a doubt, 1,100 lives were changed, and those 1,100 will touch and influence many others.

The island people's interest in having a role, an interest that had begun to build during that first eco-audit trip, blossomed into a Maui-day event. Sponsored by Interface, the hotel, and a local group called Maui 2000, the event was staged on the hotel grounds under a tent. We expected a turnout of maybe 50, but some 400 people came from all five populated islands and from as far away as Tahiti and New Zealand. Indeed, Interface's influence had already spread throughout the region. The DreamTeam spoke, one after the other—the first time this inspiring, smart, visionary, and practical team of environmental leaders had ever appeared together on the same program. How appropriate that the descendants of a people and culture that lived sustainably for 600 years in their island paradise, before we "discovered" them, should be the catalyst to bring together this modern-day equivalent of their own ancient wisdom, to inspire them to strive for sustainable living again. Maui 2000 lives on, reinvigorated and inspired by the DreamTeam.

And we have it all on tape, along with all the rest of the World Meeting. David Brower told a local interviewer that historians would look back on that day as a turning point for Earth. David said, "They [the historians] just don't know it yet." His remarks went unpublished, but someday some historian will be doing some research and David's statement will be there on our tape to be discovered, and perhaps (who can say?) the day will be marked.

The last day's program brought a moving response from the native Hawaiian community. Elders, led by 70-year-old Kapuna Kealoha, and children came together to thank us, tearfully, both in their native language and in English, for the gift of Maui Day and to receive the Children's Legacy. For them, it was an unprecedented experience with *haoles* (foreign visitors). Their thank you ceremony, in the native tradition, was touchingly sincere. Not one of the 1,100 present was unmoved, and tears flowed freely. Culturally, past and present were united. Keith McLoughlin of

DuPont, representing the suppliers, was adorned with a lei by one of the children. Keith had to bend over double to receive the tribute, so tiny was the tot. It's okay for a grown man to cry, especially at a time like that.

I was inducted into the family of Kekula Bray-Crawford, one of the community leaders who spoke to us. Kekula and I are sister, brother, aunt, uncle, mother, father, and cousins. I hope to learn what requirements that places on me when I can get back to Maui and see my sister-aunt-mother-cousin again. Of one thing I am certain: I, and any other Interface person, will be welcomed back. Meanwhile, Kekula tries to teach me by e-mail. I am learning about this old, though new to me, culture very slowly.

Legacy ideas had poured in. The DreamTeam, charged with judging and selecting the best ideas, had thrown up their hands at the difficulty of their assignment. All 40 ideas were good. How to choose? Hawken, in his inimitable and brilliant way, provided the solution: Let the Maui community, itself, choose.

Not so easy: The native community is, in fact, divided. There are factions. Some want to secede from the United States. Others want to modernize with the most advanced technologies. There are three main factions. Paul's idea was to require the factions to cooperate, each to appoint a representative to a council of elders to be formed, with the elders joined by representatives from the children of the community. A wonderfully reconciling structure, horizontal among the factions, vertical among the generations, children and elders would be brought together to think and act in cooperation and reconciling concert.

The Maui community accepted the invitation to form such a council to administer the legacy, and asked Paul and me to meet with them before we left the island. We agreed to do that, to hear their organizational plan.

The last evening brought to fulfillment what Charlie Eitel says was his initial vision for the World Meeting, dating from my first casual remark that set his thoughts and wheels in motion. It was, he says, a vision of a hooked-up Interface, all the people from those 34 countries, with key suppliers, and all the resource people who pulled the meeting together, dressed in white, all joined in a great endless circle, forming the Interface logo, the Circle I. It happened on a Maui golf course while a helicopter captured the moment. As the camera ran, a scattered crowd converged and coalesced, and the order of the Circle I emerged. It is a visual and emotional wonder, white-clad, waving people against the green backdrop of a carpet of manicured grass. The golf course with its pesticides and the helicopter with its noise were vestiges of the first industrial revolution for which we just could not find suitable substitutes. Next time, maybe.

Kenny Loggins sang for us that last evening under the stars. And Glenn Thomas sang, too. Glenn's song has special meaning, so let me tell you about it.

On a Tuesday morning in March 1996, I had talked about our environmental mission to the sales force of Bentley Mills, one of the Interface companies, during their annual sales meeting. I thought I had made a pretty good talk, but I couldn't be sure how it was received. People made nice comments, but then to me they would.

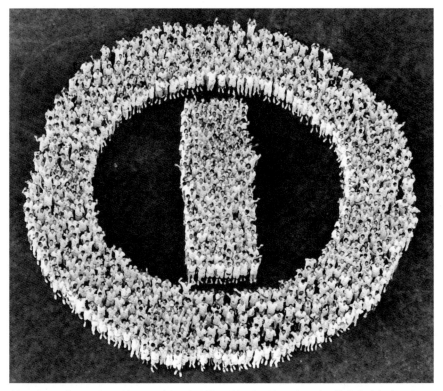

More than a thousand participants at the Maui World Meeting join to form the Circle I of the Interface, Inc. corporate logo, hooking it up.

So when a few days later, over my e-mail, totally out of the blue, came the following original poem from Glenn Thomas, it was one of the most encouraging moments of my life. It told me that at least one person in that Tuesday morning audience (and I think he surely represented many people) *really got it*. Here's what Glenn Thomas composed and sent:

Tomorrow's Child

Without a name; an unseen face
and knowing not your time nor place
Tomorrow's Child, though yet unborn,
I met you first last Tuesday morn.

A wise friend introduced us two,
and through his shining point of view
I saw a day that you would see;
a day for you, but not for me.

Knowing you has changed my thinking,
for I never had an inkling
That perhaps the things I do
might someday, somehow, threaten you.

Tomorrow's Child, my daughter-son,
I'm afraid I've just begun
To think of you and of your good,
though always having known I should.

Begin I will to weigh the cost
of what I squander; what is lost
If ever I forget that you
will someday come to live here too.

I immediately e-mailed Glenn in reply that "Tomorrow's Child" just had to be set to music, and about two weeks later, an audiotape came. It was Glenn's own voice singing "Tomorrow's Child," words and music by Glenn Thomas. By the time we got to Maui, "Tomorrow's Child" was already becoming part of our corporate culture. Glenn sang his song for 1,100 people that last evening.

Charlie's closing comments presented to the world a hooked-up Interface, and announced the World Meeting's objective accomplished! In closing, I reminded our people of Gandhi's words, which I paraphrased: "If you would have a miracle, you must be the miracle." I offered the thought that they had been there, they had been the miracle, and now should go and live it—every day of their lives, at home, at work, and at play.

We concluded this magical week together, lying on our backs in the dark but for a starlit Hawaiian sky, listening to "Clair de Lune" performed by the Maui Symphony Orchestra. Chip DeGrace, who led the staff effort to organize the meeting, chose the piece because it had been the favorite of his father, who had died the year before.

Aglow with the miracle of Maui, we began to disperse the next day. Every participant took home a videotape of excerpts from the meeting, concluding with that hook-it-up formation taking shape. Nobody's sure how our media people, Light and Power, got the tapes made overnight for every single person.

But the Hawaiian council had asked Paul and me to meet, so we did. The main topics were how to organize and what to call the council. Their first suggestion, so sensitive, so polite, was to name the council for me. As sincerely as it was offered, I took it to be a gesture. In no event could I let that happen, and I declined as gracefully as I could, suggesting they devise a better name. They were ready, and in the most considerate way offered the name: Hoòkupu. In the Hawaiian language hoòkupu has two meanings: "a gift" and "the first seed sprouting out of the devastation of a lava flow." How appropriate! It reflected their basic plan to use the more than $200,000 pledged as seed money for a non-profit organization that would, in time, with increasing funding, carry out every single idea that had been offered by the legacy teams, and more.

Searching for a way Interface could permanently be associated with the good that would flow from Hoòkupu, I told them about the origin of "Tomorrow's Child," and I recited the poem for them. The last piece fell into place as, at my request, the council adopted the name:

The Hoòkupu Trust

A Legacy for Tomorrow's Child

"Let's do something to love all the children," Bill McDonough's original vision, lives in Maui.

In a most delicious and astonishing irony, Jim Hartzfeld was on his computer days later and found a new meaning in "Hoòkupu." He removed the punctuation and spaced the letters a bit differently, and there before his eyes appeared:

Hook up u

Mission accomplished? I'll say!

Synchronicity? Or serendipity? Who can say? Which was it

when a young woman named Melissa Gildersleeve, working for the state of Washington's Department of Ecology, heard Paul Hawken speak, then bought and read his book? Which was it when she sent the book to her mother, Joyce LaValle, an Interface Regional Sales Manager in Los Angeles who was pursuing a carpet order for a project called the ERC? Which was it when Joyce, fearing that she would lose the business because the environmental consultant for the ERC, named John Picard, had said that Interface just didn't "get it," sent the book to her boss, Gordon Whitener? Which was it when Gordon sent the book on to me at a most propitious moment, just when I was sweating over what to say to a new task force to give them an environmental vision, and was trying to decide what my new role should be in the company of my own creation, *and* was trying to figure out just what John Picard meant when he said that Interface didn't "get it"? (And, yes, what Interface, itself, should grow up to be?)

Yet as powerful, galvanizing, and culminating as the Maui experience was in providing the setting in which the environmental theme could take over the process of hooking up a worldwide company, there may yet be a role of even greater importance ahead for the model that emerged from Maui. *And then what?*

In November 1996, I was appointed to the President's Council on Sustainable Development (PCSD). The PCSD had been created by President Clinton in 1993 as a council of distinguished Americans from various backgrounds and viewpoints: business people, environmentalists, labor leaders, government employees (federal, state, and local, including Cabinet members), women's activists, and Native Americans. Charged by the President to reach consensus on policy recommendations pertaining to sustainability, the members with widely divergent points of view reached an amazing degree of consensus during the first three years of the PCSD's charter. One result was a report, "Sustainable America, A New Consensus." It offers a wonderful vision of a sustainable society, one that would be difficult to fault from any point of view, especially given our starting point of where we are today, an unsustainable society.

Jonathan Lash, President of the World Resources Institute, a Washington, D.C.-based think tank, and Dave Buzzelli, Vice President of Dow Chemical Corporation, co-chaired the council through its first three years, effectively putting sustainable development on our nation's agenda, raising public awareness of the numerous issues surrounding the concept, and leading the council to reach consensus about the environmental, social equity, and economic imperatives of sustainable development. It was a remarkable achievement.

But as President Clinton approached the end of his first term, the PCSD wound down its activities, sacrificing precious momentum to the uncertainties of an election campaign. Then, with President Clinton's reelection, the Council received a charter renewal, and was charged by Vice President Gore to take on the task of implementing the vision outlined in "Sustainable America," to continue to get the word out, raise public awareness, applaud and publicize success and, in particular, take on the issue of global climate change. He asked for policy recommendations, again based on consensus, after (in the Vice President's words) "looking long, thinking big, and being creative."

Many members from the first term have left the council, and others have been appointed in their places. Co-chair Dave Buzzelli has stepped down from that position, but remains on the council, and I have been appointed co-chair to work alongside Jonathan, who continues—thank goodness.

Reluctant at first to accept Jonathan's bidding to allow him to recommend me to the White House for the co-chair, I finally relented, primarily because of something my friend Huey Johnson once said to me: "One policy is worth 10,000 programs." The Power of One, in a political sense.

I took the job and quickly found it to be a thankless task, one which I have likened to a very large sack of potatoes to be peeled—peel one and there are plenty more where that one came from. But peel on we do, and the reconstituted council, after a year, began to regain its momentum. Our first significant accomplishment was to deliver to the White House an agreed set of principles, a "this we believe" statement about global climate change. The principles are a good start, though in time we should (and, I think, will) go much further before we're finished, because Earth needs for us to. The principles are calculated to help President Clinton in climate treaty negotiations with other nations and, perhaps more importantly, with our own Senate to gain ratification of the Kyoto treaty.

Another objective, perhaps the one that can have the greatest value if we do it right, is to organize a National Town Meeting for a Sustainable America. We are planning the meeting for May 1999, hoping to make it a truly national, even international, event to stimulate a much needed phenomenon: to get America to speak up for Earth. For a network to "work," it first has to "net," to find each other and connect, to hook up (sound familiar?), so to speak. The people working on the meeting believe there is a vast potential network just waiting to "net" when people of like mind discover each other's existence and draw strength from their surprisingly large (we believe), like-minded numbers.

We are planning the conference with a center, Detroit, and with cities all over America connected by satellite television. Each major site will conduct its own concurrent local sustainability conference, alternately tuning into the center and showcasing and spotlighting local initiatives. In turn, each site will be downlinked to smaller cities, schools, universities, churches and other institutions, and homes all across America. We visualize a network of conferences to involve a new network of people, discovering each other, realizing they are not alone, and rising up with one voice to call for a sustainable society. If we're successful, business and government alike will have to listen. When the people lead, the leaders will follow.

Interface's Maui experience is the model for this renaissance, this monumental hook-it-up attempt that we hope will awaken in America the spirit and the heart of the next industrial revolution. We have a poster child for the conference, all set to go, "Tomorrow's Child." We have themes, too: The Power of One. What if everybody did it? To love the children, all the children—today's and tomorrow's—of Earth. Perhaps Tomorrow's Child will become the poster child of the next industrial revolution, and Ken Medema's song, the theme.

THE FUNDAMENTAL TRANSFORMATION of Interface, I believe, is a phenomenon of first order magnitude, the result of a synchronistic flow and confluence of events and circumstances. It is the real business school case study yet to be written, and the one most needed by business for business' sake.

Beginning at the head waters of my personal journey, framed to some unknowable extent from conception, and continuing on the path I have described, with the additions and contributions of so many other, special people, passing that milestone day of epiphany in August 1994, and incorporating all that has happened along the way, before and since, our associates and I have come to this remarkable and humbling place. We have a vision (accurate, we hope) of what might yet be, through our own actions and the power of our influence.

The reforming of our company into a diversified, global company following the collapse of the new office construction market in 1984, and the second reforming, under the leadership—the sheer operational genius—of Charlie Eitel and his management team, into a feeling and caring organization of empowered and accountable people give special character to this exceptional juncture. Charlie, in his unusually open way, exposes the secrets of that genius in his three books, *Eitel Time, Face It* (co-authored with Dr. J. Zink and me), and *Mapping Your Legacy, A Hook-It-Up Journey*—all very much worth reading.

Not the least extraordinary aspect of what we are, and the place at which we have arrived, is that we have discovered the indispensable value of the *soft* side of business—what is being referred to more and more often as *spirituality in business*. Spirituality in this context has nothing to do with religion; it is the discovery of the value in business, euphemistically speaking, of the right hemispheres of our brains, the emotional, caring, nurturing side of our personalities. Spirituality pertains to the human spirit.

According to Dr. Zink, 85 percent of men are left-brain dominant. In our male-dominant society, men have risen to the top of the corporate ladder in disproportionate numbers; so business is dominated by analytical, objective, numbers-oriented, factual, results-driven thinking. And too many of our corporate leaders do not know what they are missing. Just as a colorblind person (I know whereof I speak) has no notion of what he is missing, so it is with far too many objective, analytical, numbers-driven, seemingly unemotional male corporate leaders.

Dr. Zink recognized a left-brain dominance in me when he joined Charlie Eitel's turnaround team at Interface in 1993. Charlie himself is right-brain dominant and the most emotionally intelligent person I know. (I use *emotional intelligence* in the way Daniel Goleman has described it in his book of that title.) Dr. Zink facilitated the process of bringing Charlie and me together in a reconciliation of our different views of the world and molding us into a partnership that works remarkably well. The three of us described that experience in *Face It*. That process opened and sensitized me to those critically important aspects of emotion and feeling that emanate from the right side of the brain.

I am not overstating the case when I say that the environmental awareness revolution in our company would not have happened without that reconciliation of two different views of reality and the resulting sensitivity that opened my heart to Paul Hawken's message, beginning with the metaphorical account of the reindeer of St. Matthew Island. Further, it has been the experiential, breakthrough learning processes imparted to our people by Pecos River Learning Center and One World Learning, together with Dr. Zink's manifest concern for the care and well-being of the *children* of Interface families, that have opened our people to the unfolding process of this revolution. It follows that the destiny of our company, its higher calling, and its *ultimate strategic purpose* have flowed from that well-spring of spiritual awakening.

Just as the McDonough paradox is the reconciling synthesis of the opposing views of foot draggers and alarmists, I believe that *doing well by doing good* represents the beginning of the Hegelian synthesis of nature's interest (thesis) and business' interest (antithesis), these heretofore seemingly diametrically opposed interests that have grown out of the misbegotten first industrial revolution. In the next industrial revolution they *must* be reconciled and achieve synthesis.

The growing field of spirituality (a term that, frankly, turned me off when I first heard it, because I associated it with religiosity) in business is a cornerstone of the next industrial revolution, as I see it. I believe, too, that the ascendancy of women in business is coming just in the nick of time. It is that instinctive nurturing nature, found more frequently in women, but also present in men if they will allow it to surface, that will recognize and elevate in business the vital, indispensable role of genuine caring. Caring for human capital and natural capital (Earth) as much as we traditionally have cared for financial capital will give social equity and environmental stewardship their rightful places alongside economic progress, and move society to reinvent the means for achieving economic progress itself.

Personally, though I know intellectually the importance of achieving social equity in our company and in the world at large, I confess that I have not felt the same "spear in the chest" emotion for the issue that I have felt for the environment. I'm so very pleased that others in our top management group and throughout our company have rallied to this cause as passionately as I have to environmental sustainability. Here, too, we have a long way to go, but have begun.

We have begun when we truly value diversity among our people; when we build factories in Asia to supply Asian markets rather than to exploit Asian labor to supply home markets; when we build and operate those factories to the same high standards as our North American, European, and Australian factories; and when we install the most modern, benign, and pollution-free technology there, too, rather than obsolescent, hand-me-down machinery.

We honor social equity, and thus sustainability in its fullest sense, when we not only care about education, but also care as much about educated Thais and Chinese as we do about educated Americans. We honor social equity in the most practical way when we look for ways to provide jobs in Harlem to recycle carpet from the floors of New York City office buildings.

We honor social equity when we urge our people to become involved in their communities and pay them for the time spent in those volunteer efforts.

We honor social equity when we treat people the same everywhere and know fully and completely in our minds and hearts that each of us is an accident of birth, when health and safety in the workplace and in our communities' neighborhoods are celebrated the same the world over, and when we remember that Tomorrow's Child is black, yellow, red, and brown, as well as white.

We are beginning to get there, too, when annual sales meetings, those recharge days each year for a high-performance sales force, include spouses and children. Today's child is precious, too, and family life, so punished by long hours and frequent nights away from home for many of our people, needs to be appreciated from time to time. We have been astonished by the positive effects of Mickey Mouse and Shamu on the morale of the tough, street-wise competitors our salespeople are.

These tiny, tiny steps, taken on behalf of justice and social equity, the third leg of sustainable development (alongside economic progress and environmental restoration), are right things to do, and smart things, too. They echo the refrain, "doing well by doing good." They address the mindset behind the system, too.

Do these *soft* issues really have a place in modern business? You bet! At least the Interface management team, beginning with me and extending to Charlie Eitel and throughout our organization, thinks so. We have seen the miracle with our own eyes and participated in it. Certainly no discussion of the Interface corporate culture would be complete without it. We found spirituality at Interface before we knew what to call it.

❖ ❖ ❖ ❖ ❖

As an afterword to the Maui meeting, we were thrilled for all of our people who planned and executed it when the meeting was named Meeting of the Year by the Meeting Planners Association of America. We all knew it. I'm glad for our people that the Association recognized it, too, and bestowed their Paragon award on our team.

Finally, in a postscript to this account of our epic four-year journey thus far toward sustainability, I add this rather timely and gratifying report. *Fortune Magazine* has just named Interface, Inc. one of the 100 best companies in America to work for—based on what our own people said about the company in confidential responses to a *Fortune* survey. What better way to end—for now?

PLETSUS: Practices LEading Toward SUStainability

PEOPLE:

Customers

- Provide honest information about the known environmental impacts of your company and product
- Invite customers to audit and critique your efforts
- Share your understanding of environmental issues and natural systems with customers

Employees

Culture

- Create atmosphere that encourages employees to question status quo and take risks
- Create environment that encourages life-long learning
- Engage the creativity of all employees and associates

Understanding

- Educate all employees on the corporate sustainability vision
- Educate all employees on basic environmental principles and workings of natural systems
- Create mechanism for employees to share knowledge of best practices
- Bring in experts to address and challenge employees
- Create newsletters to report sustainability projects and challenges, including information that is not specific to the company
- Provide access to information that can help employees in their private lives, e.g., sponsor seminars on ways to save energy at home
- Have a "dumpster diving" activity to understand the makeup of your waste stream

- Use experiential learning techniques to explain complex concepts
- Hold a seminar to explain the do's and don'ts of your recycling program

Involvement

- Ask employees to give input into improving environmental impacts of their jobs
- Ask employees if there are easy or low cost things that the company could do to make their jobs more pleasant and them more productive
- Involve employees in decision making when it affects them
- Always listen to what employees have to say about issues that affect them
- Respect the knowledge and intelligence of all employees
- Create work group teams to eliminate waste in their work areas

Suppliers

- Share your corporate vision and internal framework for sustainability with suppliers
- Involve suppliers in educational opportunities to learn more about sustainability

Community

Environmental Organizations and Government Programs

- Partner with environmental organizations that work on issues important to your corporate philosophy
- Commit a percentage of profits to environmental research
- Participate in voluntary government programs with the Environmental Protection Agency, such as Green Lights, Energy Star Buildings, and Climate Wise

Networking

- Contact other companies with a similar vision; share ideas

- Work with local universities to find latest environmental technologies and understanding
- Work with universities in joint research projects
- Talk with global experts
- Search for good practices and ideas outside your company
- Share your accomplishments with others and multiply good practices through them
- Invest time and resources in organizations committed to environmental progress or sustainable development

The Public
- Develop auditing mechanisms open to public disclosure
- Make public statements in support of sustainability principles and public disclosure such as the CERES (Coalition for Environmentally Responsible Economies) Principles
- Sponsor community forums about local environmental issues
- Choose community projects to support with time and money
- Open facilities to local school children to learn about sustainability and career opportunities

Management

Corporate Strategy
- Establish top management commitment to long-term environmental strategy
- Establish corporate and divisional sustainability vision statements
- Ask for volunteers to serve as local environmental coordinators
- Establish local Green Teams to implement ideas
- Gain certification in third party assessed environmental management systems such as ISO 14001 or BS 7750
- Create a process of managing all aspects of environmental stewardship

- Develop well-defined corporate values, goals, decision making, and response mechanisms
- Evaluate product and service offerings for fit with a sustainable society

Metrics

- Measure all material and energy flows in physical and monetary units
- Develop managerial "Full Cost Accounting" system
- Audit management systems and disposal practices
- Measure material and energy flows per unit of output to adjust for changes in production levels
- Create internal "green taxes" to highlight most profitable enterprise from total cost perspective

Incentive Plans

- Give rewards to individuals or teams with the best sustainability project
- Tie monetary compensation to achieving well-defined environmental goals
- Recognize outstanding commitment and progress toward sustainability

Keeping the Enthusiasm

- Set reasonable goals and always celebrate your accomplishments
- Learn through playing games
- Develop a sense of competition and pride
- Bring in college interns to research special projects for a fresh perspective
- Volunteer for a local hands-on project as a corporate team where the results of your labor are almost immediate, e.g., plant a garden of native plants

PRODUCT:

Design

- Redesign products to use less raw materials while delivering the same or greater value
- Replace nonrenewable materials with more sustainable materials, such as:
  - Organic materials, e.g., products of nature such as wood, cotton, hemp, flax, vegetable oils, etc.
  - Organically grown and sustainable harvested materials, e.g., organic cotton and produce, certified wood products, etc.
  - Locally produced and abundant materials
  - Recycled and reclaimed post-consumer or post-industrial waste materials
  - Materials consuming lower embodied energy
- Eliminate use of hazardous chemicals
- Design products to minimize consumption of energy and auxiliary materials in use
- Design products to last longer, make products more durable
- Design products to be repaired or selectively replaced when only a portion wears out
- Develop products out of easily separated components, or out of only one material, to facilitate recycling
- Consider the entire life cycle of product, including how it will be recovered and made into another useful product

Packaging

- Design out all product packaging, e.g., "taco shell" (package is part of the product)
- Develop returnable packaging
- Deliver products in bulk
- Develop reusable packaging for work-in-process materials

- Use recycled materials
- Design packaging to be more easily recycled
- Design packaging to be safe and/or biodegradable if accidentally released into the environment

Manufacturing

Energy

DEMAND

- Reclaim waste heat from processes, furnaces, air compressors, and boilers
- Systematically review all electric motor systems to minimize installed horsepower and maximize motor efficiency
- Design pumping systems with big pipes and small motors
- Design pumping systems by laying out pipes first (to minimize distance and elbows), then motors and other equipment
- Lay out plants to minimize distance materials travel
- Research product formulations to reduce process temperature requirements
- Minimize the number of times materials are heated and/or cooled
- Install multiple small motors to handle varying volumes rather than one big motor
- Design system for expected operating conditions rather than maximum expected capacity
- Stage plant flows and energy peaks to maximize efficiency
- Use computer modeling techniques to minimize energy usage
- Research Energy Miser technology on motors
- Install power submeters on all processes to continuously monitor efficiencies

- Install automatic switches to turn off equipment at a determined time of inactivity

SOURCE

- Research and adopt alternative energy sources consistent with local surroundings, such as hydroelectric, biofuel, solar, wind power, etc.
- Negotiate Green Energy contracts with utilities
- Research soft starting/control motor technologies
- Research energy storage technologies such as flywheels

Material

- Adopt a zero waste mentality; design processes to create no waste or scrap
- Adopt a zero defect mentality; most material defects become waste
- Eliminate all smokestacks, effluent pipes, and hazardous waste
- Adopt high efficiency planning and scheduling practices to minimize waste
- Network with other companies to find waste streams that can become inputs for other processes
- Buy raw materials in bulk to minimize packaging
- Carefully segregate waste materials for reuse or recycling
- Develop processes to utilize internal scrap materials
- Develop quick stop technology to minimize waste created by off-quality processes
- Take corrective action on quality problems as far upstream as possible to minimize waste
- Closely measure all material streams to monitor material efficiency

Marketing

- Commit to taking back your products at the end of their lives
- Rent only the service component of your products, e.g., warmth and light, rather than sell the product
- Be conscious about the extent and strategy of external communications to avoid greenwash

Purchasing

Work with Suppliers

- Share your corporate purchasing policy with all suppliers
- Press suppliers to follow and document sustainable practices, and favor those that do
- Press suppliers to take back packaging or not deliver it with the product
- Buy services, not products
- Encourage suppliers to report their environmental impacts in your terms
- Encourage suppliers to develop and offer products with a smaller environmental footprint
- Ask for information about the environmental policy of the corporation and information about the specific products you buy from suppliers
- Include the waste and embodied energy used to produce raw materials purchased from suppliers in your environmental foot-print analysis

Buy Sustainably

- Establish a "Buy Sustainably" policy stating the corporate goals on specific items when possible
- Circulate a list of recycled or environmentally friendly products to purchasing staff
- Set out clear guidelines to follow
- Support training for purchasing agents to understand the issues

- Create an internal purchasing agent team focused on identifying appropriate products
- Develop environmentally responsible methods of reconditioning used products
- Share surpluses with other offices by publishing a regular list
- Implement high efficiency planning and scheduling practices to minimize waste

PLACE:

Facility

Design

- Increase insulation in walls and doors
- Use double paned or super windows
- Use high efficiency glazing or films on windows
- Use shades, deflectors, and light shelves to reduce summer sun
- Design HVAC and utility systems for maximum long-term flexibility and efficiency, e.g., under floor delivery, personal control
- Maximize use of natural ventilation heating and cooling
- Specify finishes and materials with low VOCs and that control the growth of microbial contamination
- Install fast acting doors in factory and warehouse exits to minimize time the door is open to outside
- Design minimum of impermeable surfaces to minimize storm water runoff
- Give priority to pedestrians, mass transit riders, and cyclists instead of automobile drivers
- Design with the natural flows of the site in mind
- Provide safe areas to secure bicycles
- Use minimal finishes, such as paints and coatings
- Use low embodied energy, locally abundant building materials
- Rehabilitate existing buildings rather than tear them down

- Use salvaged or refurbished materials
- Locate near existing infrastructure
- Separate and recycle construction waste

Operations

ENERGY

- Conduct an energy audit with the help of local utilities
- Replace old boilers with new high efficiency units
- Install properly sized variable speed motors/fans
- Install heat exchangers on building exhaust ducts
- Preheat boiler feedwater with waste heat
- Use excess plant heat to heat offices
- Install programmable thermostats
- Regularly maintain all HVAC systems
- Check fan speeds and efficiencies on HVAC systems
- Regularly replace filters and clean duct work
- Replace CFC in cooling systems with non ozone depleting refrigerants
- Install variable air diffusers
- Plant trees to shade eastern, western, and southern windows and air conditioners

LIGHTING

- Redesign lighting to fit work processes, resulting in productivity improvements
- Install infrared motion detectors for automatic lighting control
- Replace incandescent lighting with compact fluorescent lighting
- Retrofit existing lighting with high efficiency fluorescent or metal halide bulbs, electronic ballasts, and reflectors
- Reduce use of high bay lighting
- Maximize use of natural daylight

WATER

- Reuse water whenever possible
- Reuse boiler water
- Treat and reuse dye water
- Develop closed loops whenever possible
- Conduct water use audits, looking for leaks and waste
- Install low flow fixtures in restrooms and kitchen areas

OFFICE

PAPER

- Use recycled paper with a high percentage of post-consumer content
- Use chlorine-free paper, if available
- Use paper envelopes without windows and avoid Tyvek envelopes, so envelopes can be recycled
- Place collection containers at every work station to recycle used paper
- Reduce or eliminate paperwork and numbers of copies
- Scrutinize distribution lists
- Make copies only on request; otherwise, route material
- Maximize use of bulletin boards
- Set up copiers so that double sided copying is the norm
- Route magazines instead of getting separate copies
- Keep paper that is still good on one side (GOOS paper, Good On One Side) and make scratch pads out of it
- Communicate via e-mail when possible, and don't print your e-mail messages
- Eliminate cover sheets on faxes

ELECTRONICS

- Purchase only energy saving electronic equipment—look for the EPA's Energy Star label

- Turn off computer monitors when not in use
- Turn off your computer when you go to lunch and over night
- Use a projector instead of printing overheads for presentations
- Send used overheads back to 3M to be recycled
- Send exhausted ink jet cartridges back to their manufacturer for recycling
- Use refillable ink jet cartridges
- Lease the service of high end electronics instead of buying them (then they can be returned to the provider when you decide to upgrade instead of being disposed of)
- Buy copiers, printers, and fax machines that use refurbished parts and toner cartridges

Maintenance

- Invest in high quality maintenance to extend the life and maximize the efficiency of systems
- Use only nontoxic cleaning compounds
- Maximize use of all-purpose cleaners to reduce the number of chemicals used and to minimize potential danger of mixing
- Buy cleaners in concentrated form that can be mixed at different strengths for different purposes, reducing packaging and transportation
- Use washable mugs, glasses, plates, and utensils
- Use bulk product dispensers for beverages, condiments, etc.
- Provide convenient and easy to understand recycling centers for common waste products
- Measure all solid waste streams

Landscape

- Leave as much habitat and vegetation as possible undisturbed by construction
- Landscape to promote biological diversity

- Design to minimize impact on local environment
- Compost organic matter
- Mulch lawn clippings
- Put up bird boxes and start an employee-run nest box monitoring program
- Plant a butterfly garden near an area that employees use often
- Join the Wildlife Habitat Council
- Start an employee vegetable garden
- Create a series of nature trails for employees and their families or even for the whole community
- Xeriscape by using plants adapted to local rainfall conditions
- Use gray water to water the landscaping
- Highlight native plants that are adapted to the local environment and do not require a lot of maintenance
- Employ Integrated Pest Management to minimize use of chemical pesticides
- Install storm water retention ponds to minimize volume and temperature spikes on local waterways from rain showers
- Create bird sanctuaries in migration paths

Transportation

Product

- Ship by rail whenever possible
- Reduce weight of products to consume less energy in transport
- Favor locally produced products
- Create transportation consortiums to maximize loading of trucks with other local businesses
- Pelletize waste materials such as fiber to minimize transportation energy
- Locate facilities to minimize shipping distances to major market centers

People

- Offset employee travel and product transportation with tree planting through organizations such as Trees for Travel
- Reduce number of trips by consolidating business or through better planning
- Buy alternative fuel vehicles
- Allow employees to telecommute or work alternative hours
- Offer rebates to employees who use alternative transportation and do not consume parking space
- Offer public transportation passes to employees
- Encourage video conferencing

BIBLIOGRAPHY

Anderson, Ray, Charlie Eitel, and J. Zink. *Face It: A Spiritual Journey of Leadership*. Atlanta: The Peregrinzilla Press, 1996.

Bailey, Ronald, ed. *The True State of the Planet*. Free Press, 1995.

Bast, Joseph L., Peter J. Hill, and Richard C. Rue. *Eco-Sanity: A Common-Sense Guide to Environmentalism*. Madison Books, 1994.

Brower, David. *Let the Mountains Talk, Let the Rivers Run: A Call to Those Who Would Save the Earth*. Harper San Francisco, 1995.

Brown, Lester R., *State of the World 1998: A Worldwatch Institute Report on Progress Toward a Sustainable Society*. W.W. Norton & Company, 1998.

Brown, Lester R. *Vital Signs 1994; The Trends That are Shaping Our Future*. World Watch Institute, W.W. Norton & Company, 1994.

Brown, Lester R. *Vital Signs 1995; The Trends That are Shaping Our Future*. World Watch Institute, W. W. Norton & Company, 1995.

Brown, Lester R. *Vital Signs 1996; The Trends That are Shaping Our Future*. World Watch Institute, W. W. Norton & Company, 1996.

Brown, Lester R. *Vital Signs 1997; The Trends That are Shaping Our Future*. World Watch Institute, W. W. Norton & Company, 1997.

Capra, Fritjof. *The Turning Point: Science, Society, and the Rising Culture*. Bantam Doubleday Dell Publishers, 1988.

Carson, Rachel. *Silent Spring*. Demco Media, 1962.

Daly, Herman and John Cobb. *For the Common Good*. Beacon Publishers, 1994.

Ehrlich, Paul. *The Population Bomb*. Buccaneer Books, 1997.

Eitel, Charlie. *Eitel Time*. Atlanta: The Peregrinzilla Press, 1995.

Eitel, Charlie. *Mapping Your Legacy, A Hook-It-Up Journey*. Atlanta: The Peregrinzilla Press, 1998.

Goleman, Daniel P. *Emotional Intelligence*. Bantam Books, 1997

Gore, Al. *Earth in the Balance: Ecology and the Human Spirit*. Plume, 1993.

Hartmann, William K. and Ron Miller. *The History of Earth: An Illustrated Chronicle of an Evolving Planet*. Workman Publishing Company, 1991.

Hawken, Paul. *The Ecology of Commerce: A Declaration of Sustainability*. Harper Business, 1994.

Hawken, Paul. *The Next Economy*. Holt, Reinhart, and Winston, 1983.

Hawken, Paul, Amory Lovins and Hunter Lovins. *Natural Capitalism*. Little Brown & Company, 1999 (to be released February 1999).

Kant, Immanuel. *The Critique of Practical Reason*. 1788.

Maslow, Abraham. *Motivation and Personality*. Addison-Wesley Publishing Company, 1987.

McDonough, William. *A Centennial Sermon, Design, Ecology, Ethics and the Making of Things*, The Cathedral of St. John the Divine, New York, New York, February 7, 1993.

McDonough, William. *The William McDonough Collection, DesignTex Environmentally Intelligent Textiles*. 1995.

McDonough, William, Architects. *The Hannover Principles Design for Sustainability*. Papercraft, 1992.

Meadows, Donella H., Dennis L. Meadows, and Jorgen Randers. *Beyond the Limits: Confronting Global Collapse, Envisioning a Sustainable Future*. Chelsea Green Publishing Co., 1993.

Meadows, Donella H., "Places to Intervene in a System (in increasing order of effectiveness)", *Whole Earth*, Fall 1997, Page 78.

The President's Council on Sustainable Development. *Sustainable America - A New Consensus*. Washington, DC: The President's Council on Sustainable Development, 1996.

Quinn, Daniel. *Ishmael.* Bantam Books, 1993.

Quinn, Daniel. *My Ishmael: A Sequel.* Bantam Books, 1997.

Quinn, Daniel. *The Story of B.* Bantam Books, 1997.

Robèrt, Karl-Henrik. "Educating a Nation: The Natural Step," *In Context* *#28.* Spring 1991, Page 10.

Robèrt, Karl-Henrik, Herman Daly, Paul Hawken, and John Holmburg. "A Compass for Sustainable Development." *The Natural Step News,* No. 1, Winter 1996, Page 3.

Romm, Joseph J. *Lean and Clean Management: How to Boost Profits and Productivity by Reducing Pollution.* Kodansha, 1994.

Russell, Peter. *The Global Brain Awakens: Our Next Evolutionary Leap.* Global Brain. 1995

Stahel, Walter R. *The Product Life Factor.* The Houston Area Research Center, 1982.

Swimme, Brian. *The Universe is a Green Dragon: A Cosmic Creation Story.* Bear & Co., 1988.

Thomas, Glenn C. *Tomorrow's Child,* © 1996. All rights reserved. Reprinted with permission.

Thurow, Lester C. "Brains Power Business Growth," *USA Today.* August 18, 1997, Page 13A.